Intelligent Systems Reference Library

Volume 101

Series editors

Janusz Kacprzyk, Polish Academy of Sciences, Warsaw, Poland
e-mail: kacprzyk@ibspan.waw.pl

Lakhmi C. Jain, Bournemouth University, Fern Barrow, Poole, UK, and
University of South Australia, Adelaide, Australia
e-mail: Lakhmi.Jain@unisa.edu.au

About this Series

The aim of this series is to publish a Reference Library, including novel advances and developments in all aspects of Intelligent Systems in an easily accessible and well structured form. The series includes reference works, handbooks, compendia, textbooks, well-structured monographs, dictionaries, and encyclopedias. It contains well integrated knowledge and current information in the field of Intelligent Systems. The series covers the theory, applications, and design methods of Intelligent Systems. Virtually all disciplines such as engineering, computer science, avionics, business, e-commerce, environment, healthcare, physics and life science are included.

More information about this series at http://www.springer.com/series/8578

Yoshiteru Ishida

Self-Repair Networks

A Mechanism Design

 Springer

Yoshiteru Ishida
Department of Computer Science and
  Engineering
Toyohashi University of Technology
Toyohashi
Japan

Intelligent Systems Reference Library
ISBN 978-3-319-79955-1   ISBN 978-3-319-26447-9   (eBook)
DOI 10.1007/978-3-319-26447-9

Printed on acid-free paper

Springer International Publishing AG Switzerland is part of Springer Science+Business Media
(www.springer.com)

Preface

This book describes the struggle to introduce a mechanism that enables next-generation information systems to maintain themselves. Our generation observed the birth and growth of information systems, and the Internet in particular. To our surprise, information systems are quite different from conventional (energy, material-intensive) artificial systems, and rather resemble biological systems (information-intensive systems). Many artificial systems are designed based on (Newtonian) physics assuming that every element obeys simple and static rules; however, the experience of the Internet suggests a different way of designing where growth cannot be controlled but self-organized with autonomous and selfish agents. This book suggests using game theory, a mechanism design in particular, for designing next-generation information systems which will be self-organized by collective acts with autonomous components. Mechanism design has been studied for designing rules or protocols of economic systems based on autonomous, distributed, and selfish agents for performance. However, this design approach should be used for large-scale information systems where all the components are becoming autonomous and selfish (or in a dual perspective, in human systems where all individuals are connected and synchronized by the information systems). A mechanism in the "mechanism design" may be also defined as a "mathematical structure that models institutions through which economic activity is guided and coordinated" (Hurwicz and Reiter 2006). We are using the mechanism in this sense, although economic activity is done by exchanging resources among computers (nodes) in the information networks.

The global expansion of the Internet provides another reason for studying self-repair networks. The Internet may be analogously considered as the neural system of man–earth symbiotic systems to enhance and to synchronize global knowledge. In this analogy, information systems with sensors and actuators may be considered comparable to immune systems. This was the motivation of the previous book on immunity-based systems, which focused on mounting sensors and proposed self-recognition models. To cover the actuator part, repair actions are considered in self-repair networks.

The central theme of this book is: what happens if systems self-repair themselves? We have already explored a similar question: what happens if systems self-recognize themselves?

As an extension from reliability theory of systems involving active components of mutual (with distributed context) and self (with subject and object double-fold context of repairing and being repaired) repair, it is on a line extended from von Neumann's "Self-Reproducing Automata" (Von Neumann and Burks 1966):

"The ability of a natural organism to survive in spite of a high incidence of error (which our artificial automata are incapable of) probably requires a very high flexibility and ability of the automaton to watch itself and reorganize itself. And this probably requires a very considerable autonomy of parts."

It should be noted that von Neumann addressed also the problem of constructing reliable automata from unreliable components (involving some intrinsic error with a probability) in his lectures on "probabilistic logics and the synthesis of reliable organisms from unreliable components" (Von Neumann 1956). We have been addressing the problem: what happens if the system has such autonomy that each part can repair other parts with a certain probability?

Theoretically, self-repair networks struggle to devise a mechanism to avoid a defection-defection deadlock and all defectors ground state (from a game theoretical point of view); and to avoid locally stable attractors and to prevent the macro state of all abnormal states (absolute zero) from being globally stable (from a dynamical system point of view). In the engineering design, we focus on a new design method (model-based and simulation-based) for next-generation information systems, or artificial intelligence minding the self. This book examines the concept of applying biological closure and identity to information systems which are becoming closer year by year to biological systems, for biological systems are the ultimate reliable systems that have evolved for species, individuals, cellular, and genetic survival. We have presented several notes on applying biological closure and trials of the asymmetric approach to the modeling of biological systems:

N1. A Critical Phenomenon in a Self-repair Network by Mutual Copying (LNCS 3682, 2005)
N2. A Note on Space-Time Interplay through Generosity in a Membrane Formation with Spatial Prisoner's Dilemma (LNCS 5179, 2008)
N3. A Note on Biological Closure and Openness: A System Reliability View (LNCS 5712, 2009)
N4. A Note on Symmetry in Logic of Self-Repair: A Case of a Self-Repair Network (LNCS 6278, 2010)
N5. A Note on the Collective Identity of Indistinguishable Entities: A View from the Stable Marriage Problem (LNCS 6884, 2011)

A1 A Network Self-repair by Spatial Strategies in Spatial Prisoner's Dilemma (LNCS 3682, 2005)
A2 Asymmetric Interactions between Cooperators and Defectors for Controlling Self-repairing (LNCS 5179, 2008)

A3 Asymmetric Phenomena of Segregation and Integration in Biological Systems: A Matching Automaton (LNCS 5712, 2009)
A4 Asymmetry in Repairing and Infection: A Case of the Self-Repair Network (LNCS 6278, 2010)
A5 Asymmetric Structure between Two Sets of Adaptive Agents: An Approach Using a Matching Automaton (LNCS 6884, 2011)

This book is based on [N1, N2, and N4] for biological closure and [A1, A2, and A4] for the asymmetric approach using a probabilistic cellular automaton and (cooperative/defective) selfish agents. Among others, [N1] introduces a probabilistic cellular automaton to derive critical points of the network cleaning problem (can a network be cleaned up by mutually repairing nodes?), and [A1] introduces the spatial prisoner's dilemma approach to design a mechanism for self-repair.

We use the "self-action model" as a method to approach biological closure. Self-involvement should be avoided as it causes infinite regress, and it is hard to handle in a closed logic. However, it cannot be avoided when dealing with biological closure, for self-involvement is so intrinsic in biological systems that it provides a distinction from artificial systems. Conventional system methods view the system from outside as a collection of interacting subsystems. However, the self-action model views the system from inside as a collection of entities capable of acting on other entities and being acted upon by the other entities as well. The other entities do not exclude the case of the system itself. Also, the term "action" is related to real systems involving monitoring, diagnosing, repairing, and rearranging.

To implement the self-action model, we use agents as primitives. Thus, we implement the self-action model as a collection of agents (autonomous, distributed, and even selfish entities). We may not be able to design the self-action system like designing conventional systems beforehand and from outside, but we need to design the system as a player viewing the system from inside (i.e., from the eyes of the constituent entities: agents).

This book contains interdisciplinary research encompassing game theory (prisoner's dilemma), complex systems (probabilistic cellular automaton), reliability theory (system reliability), and particle physics (critical phenomena). The particle physics approach is taken for information networks where nodes and operations among nodes correspond to particles and interactions, respectively. For a game theoretic approach on information networks, nodes and operations, respectively, correspond to players (agents) and actions; and a mechanism design for the reliability of the entire network has been studied assuming selfish and autonomous agents. Regarding the mechanism design, which is a subfield of economics (or subfield of game theory), Myerson (Myerson 2008) defined: "Mechanism design is the fundamental mathematical methodology for analyzing economic efficiency subject to incentive constraints." Thus, the design needs to assume not only autonomous distributed agents but also selfish ones.

As a game theoretical approach to networks, while Chaps. 3 and 7–11 present a naive model based on a probabilistic cellular automaton focusing on how uniform repair affects the networks, Chaps. 4–6 introduce cooperation (repairing other agents) and defection (not repairing other agents while accepting being repaired).

Regarding the economic theory of networks, Chap. 2 is a micro theory focusing on interactions between two agents (incentive for cooperation, in particular), while other chapters examine a macro theory of emergent properties of networked agents.

References

Hurwicz, L., Reiter, S.: Designing Economic Mechanisms. Cambridge University Press (2006)

Myerson, R.B.: Perspectives on mechanism design in economic theory. Am. Econ. Rev. 586–603 (2008)

Von Neumann, J.: Probabilistic logics and the synthesis of reliable organisms from unreliable components. Automata stud. **34**, 43–98 (1956)

Von Neumann, J.V., Burks, A.W.: Theory of Self-Reproducing Automata. University of Illinois Press, Urbana and London (1966)

Acknowledgments

We are grateful to graduate students: Toshikatsu Mori, Masahiro Tokumitsu, Yuji Katsumata, Masakazu Oohashi, Kei-Ichi Tanabe, Yuta Aoki, and Yuuki Sugawara who helped conducting computer simulations; also to Yoshikazu Hata, Yuji Morita, Shigetaka Ikeno, and Shota Takagi who helped re-computation of the computer simulations.

Many thanks to Lakhmi Jain for giving an opportunity of this publication; Thomas Ditzinger for patiently waiting for my manuscript; Sankar Raj Gopalakrishnan for supporting the proofreading; and Julian Ross for revising and editing the entire draft.

I am indebted to many foundations for their financial supports. The foundations include: CASIO Science Promotion Foundation, DAIKO Foundation, NIT Foundation for the Promotion of Science, Grants-in-Aid for Scientific Research (B) 16300067, 2004; the 21st Century COE Program "Intelligent Human Sensing" of the Ministry of Education, Culture, Sports, Science and Technology of Japan; The Global COE Program "Frontiers of Intelligent Sensing", from the Ministry of Education, Culture, Sports, Science and Technology; Grants-in-Aid from Toyohashi University of Technology. Some work presented here was also supported in Grants-in-Aid for Scientific Research (B): "Foundation on Immunity-Based Complex Systems and Sensor Agents" (23300057) 2011 from Japan Society for the Promotion of Science.

I am grateful to Springer and International Journal of Innovative Computing Information and Control (IJICIC) for their permission to use figures in research papers previously published by them. Most figures appeared in a series of the lecture notes in computer science (LNCS). Figures in Chap. 9 first appeared in IJICIC. Figures in Chaps. 10 and 11 first appeared in Artificial Life and Robotics (AROB).

Finally, I am grateful to my family and my mother for supporting my academic life.

This book is written in honor of the late Herbert Alexander Simon. I would like to dedicate this book to all those who sacrificed themselves to help people attacked by the Tohoku Earthquake and Tsunami (3.11/2011); and subsequent events caused by them.

With love, respect, and thanks,

Toyohashi, Japan   Yoshiteru Ishida
August 2015

Contents

List of Figures

List of Tables

Chapter 1
Introduction: Self-Action Models

Abstract This chapter defines a general framework of self-action models, in which the self-repair network is considered. Motivations for the self-action models will be discussed by correspondence between information-intensive artificial systems (information systems involving connected computers) and biological systems. With increasing similarity between these two systems, emphasis is placed on a game theoretic approach for selfish agents and evolutionary mechanism for the autonomy and maintenance of information-intensive artificial systems. The background is also explained, introducing related paradigms of *autonomic computing, recovery oriented computing* and *grid computing*. Other related fields such as game theory (with evolutionary and spatial game theory), network sciences (with the graph theory and statistics), interacting particle systems (based on probabilistic process theory) and reliability theory are also briefly explained.

Keywords Self-action models · Autonomous distributed systems · Game theory · Mechanism design · Multi-agent systems · Complex networks · Probabilistic process theory · Reliability theory

1.1 Self-Action Models

The intrinsic features of the self-action models are as follows:

- Self-involvement;
- Autonomous and distributed systems with selfish agents;
- Asymmetry of existence and non-existence (privileged zero Kampis 1991).

If we restrict the unit of the actions and being acted upon as an agent, then the model can be neatly captured and visualized as a network where agents are

---

Some discussions of this chapter are presented in (Ishida 2008, 2007).

© Springer International Publishing Switzerland 2015
Y. Ishida, *Self-Repair Networks*, Intelligent Systems Reference Library 101,
DOI 10.1007/978-3-319-26447-9_1

1

expressed as nodes and actions as directed arcs (also called self-action networks). We use the word "action" to mean operation on the attribute related to the existence property. For example, the statement "this statement is false" is not only self-referential, but it operates on the existence property of "statement" where the existence of a (true) statement is almost always assumed. The statement of the existence property belongs to logic that is different from the one within the (symmetrical) state.

Because of the asymmetry of existence and non-existence (or asymmetry of zero and non-zero), attributes of existence are also inherited to the asymmetry in repair by normal agents and repair by abnormal agents. Examples of asymmetry between zero and non-zero can be found in the equations of population dynamics. Once the variables that express population number or density become zero, they will never recover the population and will remain zero forever.

The concept of "self-involvement" and "asymmetry of existence and non-existence" may be intuitively similar to the difficulty of generating random numbers. Let us consider devising a random number generator as a preliminary thought experiment. It may be readily assumed that order and disorder (randomness) exhibit asymmetry. Asymmetry of order and disorder is analogically similar to a linear system and nonlinear systems. A rule of order may be specified by a rule; however, a rule of disorder cannot be specified by a rule by the definition of disorder. In order to devise an algorithm, we need a specific rule to generate or even criteria to evaluate whether the generated numbers meet the specification. However, it is contradictory to imagine that random numbers can be generated by an algorithm, for the numbers generated by any algorithm must obey some generating rules or meet some criteria, which means that the generated random numbers are in fact ordered ones. For "self-involvement," this could lead to a paradox similar to the "self-maintenance paradox": Assume a random number generating algorithm $A0$ which requires one parameter $p0$ which should be a random number. Then the algorithm requires another algorithm $A1$ to feed the random number to $p0$, however, this algorithm $A1$ in turn requires a random parameter $p1$, and so forth. For ordered numbers (e.g. factorial numbers, Fibonacci numbers), the recursive way works, that is, use $A0$ instead of $A1$. However, due to asymmetry of order and disorder, the recursive procedure cannot be used and disorder (randomness) must be fed from outside of the algorithm, perhaps by an intrinsically random natural phenomenon. In practice, pseudo random numbers are generated and used by an algorithm involving seeds (a parameter fed from outside). It is controversial whether π (the ratio of the circumference of a circle to its diameter) can be used as a parameter to be fed.

Self-involvement alone logically leads to an infinite regress. Distribution can relatively relax the dichotomy of the paradox. It is indeed relative relaxation because the extent of relaxation is related to the extent of distribution. It can be traced back to von Neumann that self-expression can avoid the self-referential paradox by dividing the expression into three parts (von Neumann and Burks 1966).

Let us describe these self-involvement and solution to the paradox with an operator M and an operand S. $M(S)$ indicates the target system S with its maintenance system. The self-maintenance paradox means $S \subseteq M(S) \subseteq M(M(S))$ ... $\subseteq M^k(S) \subseteq$ ... will be monotonically increasing, hence a system that can maintain itself would not be obtained even when the system is extended an infinite number of times. To avoid the paradox, we need to devise an automaton S such that it is a fixed point of the operator M: $M(S) = S$. The approach of the self-repair network (in two nodes) amounts to the nodes being operator as well as operand: $M_2(S_1) = S_1$ and $M_1(S_2) = S_2$ where maintenance M_i is realized by the other system (node) S_j.

To focus on the interplay between self-involvement and distribution, let us consider another thought experiment of self-gambling. Suppose a scholar has a theory and does not know whether it is true or false. When the scholar has a chance of betting on whether the theory is true or not, the scholar can hedge the satisfactory level by betting that the theory is wrong, for if the theory is wrong, the scholar wins the gamble, while if the theory is correct the scholar wins in terms of academic achievement. However, the gamble cannot be made alone and the problem is whether many scholars can mutually hedge their satisfactory level by mutually betting. This distributed problem is left for readers.

Innovation related to self-involvement is self-simulation used to give a counterexample to the one-dimensional cellular automaton version of the positive rate conjecture (Gacs 2001, page 20 in Gray 2001). This is not only an example that an automaton can do actions (simulate) involving itself (self-simulate) avoiding the self-referential paradox, but also an example that self-involvement is a powerful tool for rigorous analysis and inference. Focusing on information systems, throughout this book we identify repair as a state modification, in which case a closed self-repair system is possible for such repairs. However, the practicality of realizing such systems may require further investigation, for Gacs' result is for an infinite system composed of an infinite number of automata, whereas real information systems lie between a finite system and an infinite system, that is, a growing system, and furthermore, real information systems such as the Internet are open systems (or a hybrid of human and automata).

In immunity-based systems (Ishida 2004), a *self-recognition model* (or a *self-organizing network model*) has been discussed. The self-recognition model also includes self-involvement that would require a higher level of logic. Although the "self-recognition network" inspired by Jerne's immune network has been used as an illustrative model, only one model may not suffice to understand self-involvement.

Meanwhile, network sciences have been studied extensively and several breakthroughs utilizing computer assisted statistics have been made. However, what would lead to self-involvement combined with scaling up of networks remains a mystery. Of course, fundamental concepts such as degrees, cluster coefficients and the emergence of giant clusters have played an important role in structuring and characterizing networks.

1.2   Unit in Biological Systems

The greatest difficulty in understanding biological systems (but an obvious matter for artificial systems) is the concept of unit in biological systems. Biological systems have several units: genes, cells, organs, individuals and species. At a glance, they seem to be hierarchically arranged but they are functionally entangled. There have been several discussions on unit in biological systems such as the unit of adaptation (e.g., Smith and Parker 1976; Iwasa 1987); the unit of selection (e.g., Dawkins 1989) or the object (target) of selection (e.g., Mayr 2004). In the self-action models, we need to design the scope and granularity of actions and recognitions as exemplified in Chaps. 4, 5 and 6.

A challenge for the self-action model is to extend the unit to allow functionally entangled ones like biological systems. We have just extended several scopes of actions and being acted upon to devise specific geometrical configurations to enhance the stability of collective cooperators (mutually repairing clusters protected by a membrane in Chap. 6).

The self-action models take the unit into consideration, but in a simple way: agent as a unit of actions and at the same time as a unit of receiving the actions. This approach is indeed simple, and yet involves a self-referential feature at the unit level.

It is worth mentioning that a system-level property "double-edged sword" (Chap. 3) emerges from the dual nature of agent: a unit of action and a unit of being acted upon.

Another feature of the self-action model is that "a third party" (which is able to observe the entire system objectively from outside of the system of interest) is excluded as much as possible. This nature is shared by fair and symmetric assumptions on *players* in game theory. Observations, if necessary, are assumed from each player (agent). This is in contrast to assumptions in conventional mathematical programming and conventional control theory where "a third party" (a designer) is assumed to have much information.

1.3   Causality in Biological Systems

As artificial systems become more information-intensive, causality that governs the development of artificial systems resembles that of biological systems rather than that of energy-intensive physical systems. Causality in biological systems has been extensively discussed. For example, Mayr formulated the causality in biology as *dual causality* (Mayr 1997). Here, we revisit causality in biological systems in the context of self-action models.

- Serial Causality (A cause–effect sequence forms a linear chain.)
- Emergent Causality (The higher level that emerges from the activities of the lower level controls or affects the lower level.)

- Circular Causality (A cause–effect sequence forms a cycle.)
- Evolutionary Causality (evolution of biological systems, and that of artificial systems and sciences by social transmission.)

One confusing aspect of causality is that the different types of causality look alike from a local viewpoint that one entity (cause) affects another entity (result), however, the causalities except the serial one are global (macro) causalities involving entities other than the two entities in question. A heuristic to discriminate the macro causalities from the local causality (serial causality) is that in macro causalities the cause and result can make sense even if they are switched. For example, in the "chicken or egg" problem, a chicken can be a cause of the egg and an egg can be a cause of the chicken. Other popular examples include "a sound body in a sound mind." Also, "an organized room by an organized person" could be "an organized person by an organized room."

The Internet, which may be classified as the Barabási-Albert network, which is self-similar and scale-free, may be characterized by hub nodes having a very high degree. The emergence of the hub node is explained by the high degree, and the high degree again can be a cause of even higher degree; thus, richer people become even richer.

With the experience of self-organized development of the Internet, next-generation information systems should be designed taking evolutionary causality into account. The design (or blueprint) of biological systems is implemented as DNA (genes), and is subject to change so that the phenotype can adapt to the environment.

If the similarity between biological systems and artificial information systems were to become more alike, it may be better to explicitly implement the design mechanism in the design of information systems. One possibility is to implement "strategy" (in the game theoretic framework) as genes for information systems.

1.4 Incentives for Self-Action Models

There are three main incentives for developing self-action models:

- Timed Systems: Information systems are timed systems due to causality similar to biological systems as discussed above. It is very difficult to change the framework once adopted in the early phase even if it is not an efficient one.
- Mechanism Design: Mechanism design assuming selfish agents (hence mimicking the evolutionary design of "survival of the fittest") is important before initiating deployment of next-generation information systems.
- Backup Systems: The current Internet is extremely large and growing rapidly, in spite of its weakness.

It is worth noting here that any existing systems with physical entities and functions will fail. As Murphy's law states: "If anything can go wrong, it will."

Similarly, in the context of natural disasters to which Japan is extremely prone, Terada's warning goes: "Natural disasters will return when people forget." The illusion that a system with very low failure probability will not fail comes from the failure to take time into account when defining failure probability.

The problem of mapping a probability to time is so essential that it will appear repeatedly in many forms throughout this book. However, how many times the demon should cast a dice during some specific time interval remains a challenge. In a thought experiment involving buying a lottery with the winning probability of 10^{-N}, the probability that you will not win 10^N times will be less than half (almost 0.36788). In another thought experiment, let us assume that the statement "An M8 earthquake will occur within 30 years with a probability of 0.7" was made in 2011. Does the statement remain invariant in 2015 or should it be revised? And if it should be revised, in what way? Should it be with a time of less than 26 years, or with a probability of higher than 0.7? The Markov model forgets everything that has occurred in the past (memory-less property), while the Bayesian approach focuses on the fact (or observed data) that no such earthquake has occurred during the last 4 years. But no one knows how many times the demon cast a dice during the 4 years or will do so during the coming 26 years. We can do no more than restate Terada's warning: "Natural disasters will return when people forget, and in places where people do not expect them and in ways that people do not imagine."

1.5   Selfishware and Internet Being

The global evolution of large-scale information systems such as the Internet is so chaotic that it is not only difficult to control but also hard to predict. Email systems and Web systems have expanded widely and rapidly. Many systems based on the Internet have emerged: social networking services, sensor networks, satellite networks and so on. On the one hand, information technology has been close at hand for decades in the form of workstations, personal computers, laptop computers, mobile terminals and wearable computers, while on the other, global services such as cluster computing and cloud computing have emerged.

Another feature of chaotic systems is sensitivity to the initial state, which makes it difficult to predict the future of information systems. This suggests the importance of guiding future information systems by a mechanism design whereby the fundamental process of development is built in to regulate evolution in the space-time framework. (Even chaotic systems follow physical causality which can be determined based on the rules for changes and the initial conditions; in contrast, in biological causality not only the state but also the rules change, and causality is subject to the asymmetric feature of existence logic.)

In the microscopic model, we have shown the possibility of cooperation emerging when a unit of receiving payoff broadens such as system reliability and further availability is taken into account, rather than simply counting the cost of repair. Indeed, when the payoff is modified to a more systemic one involving neighbors' resources in the macroscopic model, a cooperative strategy that will support neighbors by repairing can exist even when a defective strategy exists. Further, even adaptability to the environment (such as fault probability and maximum resources) can emerge.

Strategic repair cannot outperform repair that is optimized to the environment, however, the optimal repair changes when the environment changes. Although strategic repair is not the best in any environment, it has shown reasonably sound performance in any environment. This is due to the fact that strategic repair amounts to distributing the repair rate in a spatiotemporal sense: distinct agents can have distinct repair rates and further each agent can have distinct repair rates at distinct times. The merit of spatiotemporal flexibility of strategic repair will be more conspicuous in spatiotemporally dynamic environments: e.g. the failure rate could vary from agent to agent and from time to time.

Since the Internet has become so large that it now resembles an ecosystem, let us consider the entity identified as a species (in an analogical sense) with two types: (cooperative) internet being and (defective) selfishware.

Although we focus on self-repairing tasks, the tendency would hold for other tasks that require cooperation among agents. The condition for the existence of both selfishware and internet being is obviously that all the players involved in the entity will benefit from it or at least do not suffer from it. For the existence and maintenance of internet being, however, a further condition is needed: payoff involves not only the short-sighted one but also a more systemic and longer term basis.

This explains two phenomena observed in the Internet: that hypertexts (web documents) are posted and linked so explosively, and that computer viruses, worms, spyware and spam mail cannot be eliminated. The former is considered to be an example of internet being, corresponding to an organization of cooperators, while the latter is an example of selfishware but not internet being, corresponding to a spatiotemporally isolated cluster of defectors (hence exploiting neighbors).

The linked network of documents (hypertexts or Web) and social networking services (SNS such as facebook, twitter and Pinterest) benefit all the players involved, that is not only the readers but also the providers (the ones who make posts), thus forming a Nash equilibrium (Nash 1950). Viruses and spam mail benefit only the providers while harming the users; users have to pay a high cost to eradicate them and instead choose to neglect them, again forming a Nash equilibrium.

In contrast to spam mail, the usual e-mail network is the internet being in the sense that both senders and receivers benefit from it. Although spam mail can appear as "intruders," it cannot destroy the usual e-mail service; the usual e-mail communication is an evolutionarily stable strategy (ESS) (Maynard Smith 1982).

A game theoretic study on complex systems such as the Internet will reveal that the network is a "culture media" for the emergence of mutually supporting collectives, since it allows internet beings to emerge when certain conditions are met. As the network is used as a repository of knowledge and data, the network can be a concentrator of computational intelligence when an organization such as grid computing is in operation. Then parasitic computing (Barabási et al. 2001) will also emerge.

1.6   Related Paradigms and Fields

The self-repair network is a model to simulate and analyze autonomous distributed systems which share the approach of computing paradigms such as *autonomic computing* (Horn 2001; Kephart and Chess 2003; Murch 2004), *recovery oriented computing* (Brown and Patterson 2001; Brown 2004; Patterson et al. 2005); *grid computing* (Foster and Kesselman 2003; Foster et al. 2008) and *cloud computing* (Foster et al. 2008) as an infrastructure that frees humans from the recovery and repair tasks.

One marked character of self-action models (hence, self-repair networks) is self-involvement. Therefore, self-repair networks also share self-star properties (such as self-organization and self-awareness Babaoglu et al. 2005) in common with the models discussed in the self-star properties. One difference from them may be that self-repair networks use self-involvement as a method to build a model for automatic and autonomous self-maintenance systems.

The project of *self-aware computing* (Hoffmann et al. 2010) also involves self-actions based on sensor values measured at several components of computers. The self-actions in self-recognition computers are based on control theory to control the variables and programmable states of the components.

Embryonic electronics (Embryonics projects de Garis 1993), which learn self-replicating and self-repairing mechanisms from developmental biology, motivated studies on the self-repairing mechanism of cicatrization in an extended cellular automata model (named multi-cellular automata with cellular division and cellular differentiation Stauffer et al. 2006) and implemented on field-programmable gate arrays (FPGA) (Mange et al. 1997, 2000).

For memory devices such as random access memory (RAM), built-in self-repair (BISR) has been extensively studied (Huang et al. 2007).

For networks, self-repairing is involved in many levels: peer to peer networks (Kuhn et al. 2005), routing in ad hoc networks (Mottola et al. 2008), multi-agent networks (De Pinninck et al. 2010) and graph theoretical studies on minimal self-repairing graphs which are immune to failure and maintain the function of transferring messages without delay (Farley and Proskurowski 1993, 1997).

Motivated by the problem of introducing self-repairing in nanostructures, self-repairing biomolecular systems have been proposed based on reassembly of DNA tiles in DNA lattices (Majumder et al. 2006). Many materials such as plastics (Takeda et al. 2003), polymeric materials (Niu et al. 2005; Colquhoun 2012) have been studied and developed for active protection or autonomic repair including self-repairing coating technology (Shchukin and Möhwald 2007; Schmets et al. 2007).

In pursuit of self-assembly from locally-interacting homogenous agents (Nagpal 2002; Klavins 2007) (amorphous computing Abelson et al. 2000; botanical computing Abelson et al. 2000), self-repairing topologies have been studied (Clement and Nagpal 2003).

As for mechanical systems, self-assembly and self-repairing modular machines have been studied (Murata et al. 2001); and a scale independent self-reconfiguration mechanism proposed, thus allowing self-repair when the modules are removed (Stoy and Nagpal 2004).

For self-repair (and self-assembly also) of many of these systems such as mechanical, material and cellular computing, spatial dynamics plays an important role. Among others, spatial (or spatiotemporal) equilibrium and stability are important concepts, and will be explored in Chap. 6 where a membrane is formed and the membrane separates the internal cluster of cooperators from the outer defectors.

Two distinct computational methods are used throughout this book instead of mathematical elaboration: agent-based simulations and numerical computation of steady state with a mean field approximation.

Agent-based simulation is a simulation technique realizing a virtual society or network where the active players are modeled as agents. Agents are atoms in programming the simulator where they can be collections of individualized and hence heterogeneous players, or simpler homogeneous players.

1.6.1 Game Theory (with Evolutionary and Spatial Game Theory)

Game theory (von Neumann and Morgenstern 1944) dramatically changed the way of system theory, and design theory in particular. The invention of game theory may be comparable to the invention of heliocentric (Copernican) theory in the age of geocentric (Ptolemaic) theory, or relativity theory in the age of Newton's absolute theory. Game theory allows multiple agents to be fairly treated; everything assumed for one agent must be assumed for other agents, thus allowing a unique system theory with a changing environment (for one agent, other agents are components of the environment) with symmetric structure (agent exchange symmetry).

Game theory is further advanced by the invention of the Nash equilibrium (Nash 1950a, b), which implies that an agent-based optimization may lead to deadlocks. Thus the Nash equilibrium may be characterized as agent-wise local minima:

- Non-cooperative equilibrium
- Deadlock of the best responses by selfish agents

Due to its non-cooperative nature, the Nash equilibrium may not be stable. Hence, the evolutionarily stable strategy (ESS) (Maynard Smith 1982) in evolutionary game theory is characterized by a narrower condition implying the stability of the strategy among specified strategies. A strategy is said to be ESS if a mutant strategy could not invade the strategy. In the invasion process, the mutant strategy needs to interact with the original strategy.

In our struggle to design a protection mechanism of collective cooperators with a specific geometric shape, we need to extend the above equilibrium and stability called spatial equilibrium and stability. In the spatial equilibrium, two strategies with specific spatial configurations (shapes) are in equilibrium due to the specific shape. If the shape undergoes a small change, the shape will recover in the spatially stable case and will collapse otherwise (spatially unstable case). We will observe that a membrane around a cooperative cluster forms a symmetrical shape (Chap. 6) like a soap bubble forming a spherical shape in a symmetric boundary condition.

Prisoner's dilemma (Axelrod and Hamilton 1981) in game theory is used in Chap. 2 to explain the incentive for cooperation between two nodes, which amounts to a microscopic model (similarly to microeconomics). Concepts of reliability theory such as reliability and availability are also involved with a probabilistic model of the Markov model.

Spatial prisoner's dilemma (Nowak and May 1992) is used in Chap. 4 to regulate repairing among nodes, which amounts to a mechanism design to maintain the self-repair network. The dichotomy of cooperation and defection in the prisoner's dilemma corresponds respectively to repairing and not repairing other nodes.

As a game theoretical approach to networks, while Chaps. 3 and 7–11 present a naive model based on a probabilistic cellular automaton focusing on how uniform repairs affect networks, Chaps. 4–6 introduce cooperation (repairing other agents) and defection (not repairing other agents while accepting being repaired).

In terms of the economic theory of networks, Chap. 2 is a micro theory focusing on interactions between two agents (incentive for cooperation, in particular), while other chapters concern a macro theory of emergent properties of the entire network of agents.

1.6.2   Network Science (with Graph Theory and Statistics)

A new way of viewing networks from topological or structural viewpoints has been proposed. Many artificial and natural networks have been studied from the viewpoint and their structural properties such as "scale-free" (Barabási and Frangos 2002) or "small-world" (Watts 1999, 2004) have been examined. On the other hand, studies on the interaction in the network or "dynamics" of the network (as opposed to structure) are also important. For example, information propagation in a "scale-free" network seems to be important. Statistical physics approaches to network science have been done for information systems such as the Internet (Pastor-Satorras and Vespignani 2007) including even immunization of complex networks (Pastor-Satorras and Vespignani 2002).

In the context of fault-diagnosis, the impact of the "scale-free" network theory is that the Internet is robust against random faults but vulnerable to selective attacks targeting the hub (Dezso and Barabási 2002), and that viruses and worms may not be eradicated since the "epidemic threshold" is absent in the "scale-free" network (May and Lloyd 2001; Pastor-Satorras and Vespignani 2001).

Another point is that malicious faults such as viruses and worms completely change the nature of faults. Machine faults are usually random faults, since they do not occur "selectively," and also they correspond to designed functions and hence are recovered when the parts covering the function are replaced. In contrast, viruses and worms change the intended function rather than causing malfunction, and hence cannot be treated by usual redundancy and/or stand-by techniques. This drastic change of the aspect of faults calls for the design paradigm of the immune system which has the "self" that is adaptive to the environment.

We have been studying the interaction of the network and proposed a self-recognition model based on the immune network to increase robustness and adaptability to the dynamic environment (Ishida 2004). In the framework, the network is double-sided and would raise the self-referential paradox in a flat logic without distribution, and hence subtle tuning is needed as in the immune system. Organizing responses to faults in large-scale systems has been proposed in a new approach of recovery-oriented computing (Brown and Patterson 2001).

For physical systems such as mechanical systems, they are repaired by identifying the faulty components and replacing them with non-faulty ones. For information systems, however, they can be repaired by simply copying the clean system to the contaminated system. As a first step toward the adaptive defense of information systems, we consider self-repairing of the network by mutual copying (Ishida 2005).

In self-repair networks with autonomous distributed nodes (which may be called agents when the emphasis is on autonomy and selfishness), abnormal nodes may adversely affect the system when they try to repair other normal nodes. Thus, the problem involves the double-edged sword similar to the immune system. We consider a critical phenomenon by modeling the problem by a probabilistic cellular automaton (pCA) which turns out to be a generalization of the well-known model (Domany and Kinzel 1984). Self-repairing cellular automata have been attracting attention in statistical physics (Gacs 2001).

Since the problem involves the double-edged sword leading to a critical phenomenon, repairs have to be decided giving consideration to the resources used and remaining in the system, and the network environment. When the self-repair is done in an autonomous distributed manner, each node does not voluntarily repair other nodes to save its own resources, thus leaving many abnormal nodes not repaired. This situation is similar to the dilemma that would occur in the Prisoner's Dilemma (Axelrod 1987, 1984). Thus, we use a spatial version of the Prisoner's Dilemma (Nowak and May 1992; Nowak 2006) for the emergence of cooperative collectives and for controlling copying to save resources.

As discussed in the previous section, the approach of self-involvement inevitably leads to the double-edged sword. The problem of cleaning a network by mutual copying is formalized, and the problem also has the aspect of double-edged sword. Due to this aspect, there could be critical phenomena that occur at the boundary of eradication of abnormal nodes. Chapter 3, "A Phase Transition in Self-Repair" uses a probabilistic cellular automaton to analyze the critical phenomena. The double-edged sword requires careful regulation. Chapter 4, "Controlling Repairing

Strategy" introduces a game theoretical approach involving the spatial Prisoner's Dilemma for the regulation.

The adaptive capability of the mechanism with the dichotomy of repair/not repair is demonstrated in Chap. 5 in a dynamic context where the environment changes spatiotemporally. An evolutionary mechanism of selecting the fittest strategy allows the mechanism to adapt to the dynamic environment if the speed of change is limited.

Chapter 6 further demonstrates the protective function (of cooperation cluster) by forming a membrane in a (spatiotemporally) generalized framework of the spatial prisoner's dilemma. Spatial strategy (as opposed to conventional temporal strategy) which amounts to a spatial version of tit-for-tat (TFT) is designed to form a membrane with a spatial version of generosity.

1.6.3 Interacting Particle Systems (Based on Probabilistic Process Theory)

Studies on interacting particle systems (Liggett 1985, 1999) and percolation theory (Durrett 1988; Grimmett 1989; Stauffer and Aharony 1994) have investigated nonlinear phenomena such as a (second order) phase transition (Konno 1994) based on mathematical models with probabilistic investigations, and have challenged the problem of universality. The theory derives critical points by mathematical analysis or by extensive computer simulations and numerical analysis. Technically, the threshold probability when the cluster of infinite size appears ("percolation") is calculated. We also use this technique partially in Chaps. 3 and 9.

The giant cluster of normal nodes may be used as a base of issuing chain repairs, or the cluster of abnormal nodes may be split into several separated clusters, however, the significance of the percolation requires further elaboration. A probabilistic model with (changeable) state such as the contact process, and the *abcde* system (Sudbury 1998) in particular, is also worth considering in relation to self-repair networks.

Self-repair networks deal with a "percolation" of normal nodes, and hence are related to the theory of percolation. As a model, the self-repair network can be equated with the Ising model through probabilistic cellular automata (Chap. 3). The self-repair networks can be also related to a specific epidemic model: the SIS model in a certain limited case of parameters (Chap. 10). One of the probabilistic models called the contact process (Konno 1994; Liggett 1999) when applied to a network (Masuda and Konno 2006) can also be regarded as a (spatially) detailed version of the SIS model.

A probabilistic approach similar to the above (critical phenomena) is used in Chap. 3 to derive thresholds when many nodes interact (repair) mutually, which amounts to a macroscopic model (similarly to macroeconomics). Concepts of complex systems such as probabilistic cellular automata (equated to the Ising Model) are used.

1.6.4   Reliability Theory

Reliability theory (e.g., Shooman 1968; Barlow and Proschan 1975) offers structures where component reliabilities are reflected in the system-level reliability depending on the structure of how components are functionally related. The system-level reliability is expressed as a function (called *structure function*) of component reliability. The system is called a *coherent system* when the structure function satisfies two conditions: monotonicity (the system reliability of a system A is not greater than that of a system B of the same structure if the faulty set in system A includes that in system B) and boundary conditions (a system is fully reliable if all the components are fully reliable; and a system is faulty if all the components are faulty).

Reliability theory has developed not only several structures such as serial, parallel, threshold (m out of n) and Boolean logic, but also systems with repair systems where faulty components can be repaired. For self-contained autonomous systems such as some information systems, components should have a double-sided character: a node of being repaired and repairing. We call such components with actions based on possibly selfish decision "agents." Chapter 2 tries to extend the structure with agents based on the fundamental concept of reliability and the possibility of its extension involving a game theoretic viewpoint.

It should be noted, however, that a model of a self-diagnosable system (which also has a double-sided character of being diagnosed and diagnosing) has been proposed (Forbes et al. 1965; Preparata et al. 1967; Ramamoorthy 1967) and studied extensively in several communities such as fault-tolerant computing (e.g., Pradhan 1980; Nelson 1990) and dependable computing (e.g., Laprie 1985, 1995).

Logics involving AND-repair and OR-repair is explored in Chap. 7. Expressions of parameters exhibit several symmetries including AND-OR duality. Approximation techniques such as mean field approximation are used to derive parameters in the steady state.

1.7   Structure of the Following Chapters

This book is organized as follows (Fig. 1.1): Chapter 2 explains the motivations for self-repair or the incentive for selfish agents for mutual repair with two agents capable of repairing others. The model amounts to an extension of reliability theory. Chapter 3 presents a result of critical phenomena in the network cleaning problem: can a network clean itself by mutual copying? The results include specific values of thresholds in several protocols of repairing. The model uses a probabilistic cellular automaton essentially equivalent to the Ising model (Domany-Kinzel model). Chapter 4 is an illustrative example of the mechanism design of the self-repair strategy. Involving a concept of spatial strategy, the network not only cleans itself up but also adapts to dynamical environments. Chapter 5 further investigates the adaptive capability. Chapter 6 studies the emergence of a membrane that would help protect *cooperating*

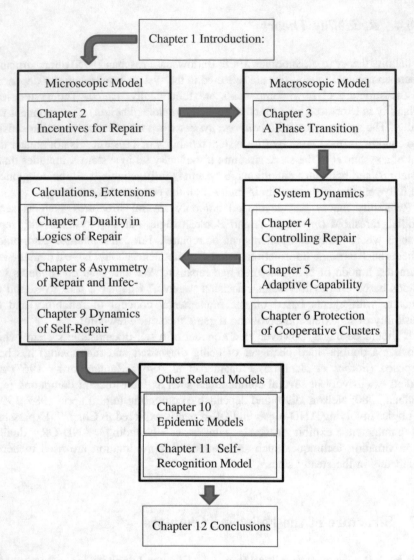

Fig. 1.1 Structure of chapters

agent clusters from being contaminated by *defecting* agents. As a mathematical model of a membrane, it is a nonlinear system having a threshold where the threshold can be identified by a mathematical elaboration. Chapter 7 summarizes thresholds in several cases. Chapter 8 focuses on how the topology of the network, self and mutual repair in particular, affects the thresholds. Infection is also included to observe the asymmetric interaction between infection and repair. Chapter 9 explains a possible application of the self-repair network to the antivirus systems of the network.

This book draws a global picture through a mapping between different disciplines. For example, the self-repair network is related to reliability theory (coherent systems) in reliability engineering (Chap. 2); the Domany-Kinzel model in statistical physics (Chap. 3); the Prisoner's Dilemma in game theory (Chap. 4); adaptive and distributed control in control theory (Chap. 5); spatiotemporally extended Prisoner's Dilemma (Chap. 6); duality in logics (Chap. 7); epidemic models (Chaps. 8 and 10); and self-recognition models (Chap. 11). In particular, a self-repair network is mapped to: the Domany-Kinzel model in statistical physics (Chap. 3); duality in logics (Chap. 7); and epidemic models (Chap. 10).

Dynamics including infection is studied in Chap. 8 which uses not only the mean field approximation but also computer simulations. Dynamics focusing on transient states in Chap. 9 relies on computer simulations with phase diagrams tailored for the transient states.

Chapter 10 studies the relation between self-repair networks and epidemic models. It is shown that the self-repair network is reduced to the SIS model in a restricted situation.

Chapter 11 discusses the relation and integration of the self-repair network and the self-recognition model, particularly the relation between the repair success rate and the recognition success rate.

Chapter 12 concludes with possible theoretical extension and implications for the design of information systems.

Throughout the book, each chapter focuses on a specific problem (written within a box). Since each chapter deals with a distinct model, the model assumptions are also boxed. In the first five chapters, models of the self-repair networks are discussed. In Chaps. 6–9, some aspect of the self-repair network is focused. In Chaps. 10 and 11, the self-repair network is related to the models of other domains. The following are the first boxes of the problem and the model assumption:

The problem throughout the book is the network cleaning problem:

> Can a network clean itself by mutual copying?

The model assumptions throughout the book are:

> - Fault: Each node has a binary state; i.e. normal (0) or abnormal (1).
> - Repair: Repair is always the interaction among the networked nodes.

The book aims to introduce self-repair networks as a model of autonomous information systems as well as self-actions as a method to model autonomous systems. More specifically, we place much weight on spatial development in addition to temporal development as in conventional dynamical systems. Several spatiotemporal extensions of concepts such as spatial strategy, spatial generosity (Chap. 4), spatial equilibrium and stability (Chap. 6) are discussed.

References

Abelson, H., Allen, D., Coore, D., Hanson, C., Homsy, G., Knight Jr, T.F., Nagpal, R., Rauch, E., Sussman, G.J., Weiss, R.: Amorphous computing. Commun. ACM **43**(5), 74–82 (2000)

Axelrod, R.: The Evolution of Cooperation. Basic Books, New York (1984)

Axelrod, R.: The evolution of strategies in the iterated Prisoner's dilemma. The dynamics of norms, pp. 199–220 (1987)

Axelrod, R., Hamilton, W.D.: The evolution of cooperation. Science **211**(4489), 1390–1396 (1981)

Babaoglu, O., Jelasity, M., Montresor, A., Fetzer, C., Leonardi, S., van Moorsel, A., van Steen, M.: The self-star vision. In: Self-star Properties in Complex Information Systems, pp. 1–20. Springer, New York (2005)

Barabási, A.-L., Frangos, J.: Linked: The New Science of Networks Science of Networks. Basic Books (2002)

Barabási, A.-L., Freeh, V.W., Jeong, H.W., Brockman, J.B.: Parasitic computing. Nature **412** (6850), 894–897 (2001). doi:10.1038/35091039

Barlow, R.E., Proschan, F.: Statistical theory of reliability and life testing: probability models. In: DTIC Document (1975)

Brown, A.: Recovery-oriented computing: Building multitier dependability (2004)

Brown, A., Patterson, D.A.: Embracing failure: a case for recovery-oriented computing (ROC). In: High Performance Transaction Processing Symposium 2001, pp. 3–8

Clement, L., Nagpal, R.: Self-assembly and self-repairing topologies. In: Workshop on Adaptability in Multi-Agent Systems, RoboCup Australian Open 2003

Colquhoun, H.M.: Self-repairing polymers: materials that heal themselves. Nat. Chem. **4**(6), 435–436 (2012)

Dawkins, R.: The Selfish Gene, 1976, Revised edition. Oxford University Press, Oxford (1989)

de Garis, H.: Evolvable hardware genetic programming of a Darwin machine. In: Artificial Neural Nets and Genetic Algorithms, pp. 441–449. Springer, New York (1993)

De Pinninck, A.P., Sierra, C., Schorlemmer, M.: A multiagent network for peer norm enforcement. Auton. Agent. Multi-Agent Syst. **21**(3), 397–424 (2010)

Dezso, Z., Barabási, A.L.: Halting viruses in scale-free networks. Phys Rev E **65**(5) (2002). doi:10.1103/Physreve.65.055103

Domany, E., Kinzel, W.: Equivalence of cellular automata to Ising models and directed percolation. Phys. Rev. Lett. **53**(4), 311–314 (1984)

Durrett, R.: Lecture Notes on Particle Systems and Percolation. Wadsworth & Brooks/Cole Advanced Books & Software (1988)

Farley, A.M., Proskurowski, A.: Self-repairing networks. Parallel Process. Lett. **3**(04), 381–391 (1993)

Farley, A.M., Proskurowski, A.: Minimum self-repairing graphs. Graphs Comb. Asian J. **13**(4), 345–352 (1997)

Forbes, R., Rutherford, D., Stieglitz, C., Tung, L.: A self-diagnosable computer. In: Proceedings of the Fall Joint Computer Conference, 30 Nov–1 Dec 1965, Part I, pp. 1073–1086. ACM

Foster, I., Kesselman, C.: The Grid 2: Blueprint for a New Computing Infrastructure. Morgan Kaufmann, San Francisco (2003)

Foster, I., Zhao, Y., Raicu, I., Lu, S.: Cloud computing and grid computing 360-degree compared. In: Grid Computing Environments Workshop, 2008. GCE'08, pp. 1–10. IEEE Grid Computing

Gacs, P.: Reliable cellular automata with self-organization. J. Stat. Phys. **103**(1–2), 45–267 (2001). doi:10.1023/A:1004823720305

Gray, L.F.: A reader's guide to Gacs's "Positive Rates" paper. J. Stat. Phys. **103**(1–2), 1–44 (2001). doi:10.1023/A:1004824203467

Grimmett, G.: Percolation. Springer, New York (1989)

Hoffmann, H., Maggio, M., Santambrogio, M.D., Leva, A., Agarwal, A.: SEEC: A Framework for Self-aware Computing (2010)

Horn, P.: Autonomic Computing: IBM's Perspective on the State of Information Technology (2001)

Huang, R.-F., Chen, C.-H., Wu, C.-W.: Economic aspects of memory built-in self-repair. IEEE Des. Test **24**(2), 164–172 (2007)

Ishida, Y.: A critical phenomenon in a self-repair network by mutual copying. In: Knowledge-Based Intelligent Information and Engineering Systems, pp. 86–92. Springer, New York (2005)

Ishida, Y.: Complex systems paradigms for integrating intelligent systems: a game theoretic approach. In: Computational Intelligence: A Compendium, pp. 155–181. Springer, New York (2008)

Ishida, Y.: Immunity-Based Systems: A Design Perspective. Springer, New York (2004)

Ishida, Y.: Information Networks as Complex Systems: A Self-Repair and Regulation Model (2007)

Iwasa, Y.: Optimal effort distribution of foraging animals. In: International Symposium on Mathematical Topics in Biology, Kyoto, Japan, 11–12 Sept 1987, pp. 49–58

Kampis, G.: Self-Modifying Systems in Biology and Cognitive Science. Pergamon Press, Oxford (1991)

Kephart, J.O., Chess, D.M.: The vision of autonomic computing. Computer **36**(1), 41–50 (2003)

Klavins, E.: Programmable self-assembly. IEEE Control Syst. **27**(4), 43–56 (2007)

Konno, N.: Phase Transitions of Interacting Particle Systems. World Scientific, Singapore (1994)

Kuhn, F., Schmid, S., Wattenhofer, R.: A self-repairing peer-to-peer system resilient to dynamic adversarial churn. In: Peer-to-Peer Systems IV, pp. 13–23. Springer, New York (2005)

Laprie, J.-C.: Dependable computing and fault-tolerance. Digest of Papers FTCS-15, pp. 2–11 (1985)

Laprie, J.-C.: Dependable computing: concepts, limits, challenges. In: FTCS-25, the 25th IEEE International Symposium on Fault-Tolerant Computing-Special Issue 1995, pp. 42–54

Liggett, T.M.: Stochastic Interacting Systems: Voter, Contact and Exclusion Processes. Springer, Berlin (1999)

Liggett, T.M.: Interacting Particle Systems (1985)

Majumder, U., Sahu, S., LaBean, T.H., Reif, J.H.: Design and simulation of self-repairing DNA lattices. In: DNA Computing. pp. 195–214. Springer, New York (2006)

Mange, D., Sipper, M., Stauffer, A., Tempesti, G.: Toward self-repairing and self-replicating hardware: the embryonics approach. In: The Second NASA/DoD Workshop on 2000 Evolvable Hardware, 2000. Proceedings, pp. 205–214. IEEE

Mange, D., Madon, D., Stauffer, A., Tempesti, G.: Von Neumann revisited: a turing machine with self-repair and self-reproduction properties. Robot. Auton. Syst. **22**(1), 35–58 (1997)

Masuda, N., Konno, N.: Multi-state epidemic processes on complex networks. J. Theor. Biol. **243**(1), 64–75 (2006)

May, R.M., Lloyd, A.L.: Infection dynamics on scale-free networks. Phys. Rev. E **64**(6), 066112 (2001)

Maynard Smith, J.: Evolution and the Theory of Games. Cambridge University Press, Cambridge, New York (1982)

Mayr, E.: This is Biology: The Science of the Living World. Universities Press, Hyderabad (1997)

Mayr, E.: What Makes Biology Unique?: Considerations on the Autonomy of a Scientific Siscipline. Cambridge University Press, Cambridge (2004)

Mottola, L., Cugola, G., Picco, G.P.: A self-repairing tree topology enabling content-based routing in mobile ad hoc networks. IEEE Trans. Mobile Comput. **7**(8), 946–960 (2008)

Murata, S., Yoshida, E., Kurokawa, H., Tomita, K., Kokaji, S.: Self-repairing mechanical systems. Auton. Robots **10**(1), 7–21 (2001)

Murch, R.: Autonomic Computing. IBM Press, New Jersey (2004)

Nagpal, R.: Programmable self-assembly using biologically-inspired multiagent control. In: Proceedings of the First International Joint Conference on Autonomous Agents and Multiagent Systems: Part 1, 2002, pp. 418–425. ACM

Nash, J.F.: Equilibrium points in n-person games. Proc. Natl. Acad. Sci. **36**(1), 48–49 (1950a)

Nash, J.F.: The bargaining problem. Econom.: J. Econom. Soc. 155–162 (1950b)

Nelson, V.P.: Fault-tolerant computing: fundamental concepts. Computer **23**(7), 19–25 (1990)

Niu, W., O'Sullivan, C., Rambo, B.M., Smith, M.D., Lavigne, J.J.: Self-repairing polymers: poly (dioxaborolane)s containing trigonal planar boron. Chem. Commun. **34**, 4342–4344 (2005)

Nowak, M.A.: Evolutionary Dynamics: Exploring the Equations of Life. Harvard University Press, Cambridge (2006)

Nowak, M.A., May, R.M.: Evolutionary games and spatial chaos. Nature **359**(6398), 826–829 (1992)

Pastor-Satorras, R., Vespignani, A.: Evolution and Structure of the Internet: A Statistical Physics Approach. Cambridge University Press, Cambridge (2007)

Pastor-Satorras, R., Vespignani, A.: Epidemic spreading in scale-free networks. Phys. Rev. Lett. **86**(14), 3200–3203 (2001)

Pastor-Satorras, R., Vespignani, A.: Immunization of complex networks. Phys. Rev. E **65**(3), 036104 (2002)

Patterson, D., Brown, A., Fox, A.: Recovery Oriented Computing. Berkeley, CA (2005)

Pradhan, D.: Fault-tolerant computing. Computer **13**(3), 6–7 (1980)

Preparata, F.P., Metze, G., Chien, R.T.: On the connection assignment problem of diagnosable systems. IEEE Trans. Electron. Comput. **6**, 848–854 (1967)

Ramamoorthy, C.: A structural theory of machine diagnosis. In: Proceedings of the Spring Joint Computer Conference, 18–20 Apr 1967, pp. 743–756. ACM

Schmets, A.J., van der Zaken, G., Zwaag, S.: Self Healing Materials: An Alternative Approach to 20 Centuries of Materials Science, vol. 100. Springer, New York (2007)

Shchukin, D.G., Möhwald, H.: Self-repairing coatings containing active nanoreservoirs. Small **3** (6), 926–943 (2007)

Shooman, M.L.: Probabilistic Reliability: An Engineering Approach, vol. 968. McGraw-Hill, New York (1968)

Smith, J.M., Parker, G.A.: The logic of asymmetric contests. Anim. Behav. **24**(1), 159–175 (1976)

Stauffer, D., Aharony, A.: Introduction to Percolation Theory. Taylor and Francis, London (1994)

Stauffer, A., Mange, D., Tempesti, G.: Bio-inspired computing machines with self-repair mechanisms. In: Biologically Inspired Approaches to Advanced Information Technology, pp. 128–140. Springer, New York (2006)

Stoy, K., Nagpal, R.: Self-repair through scale independent self-reconfiguration. In: 2004 IEEE/RSJ International Conference on Intelligent Robots and Systems, 2004 (IROS 2004). Proceedings, pp. 2062–2067. IEEE

Sudbury, A.: A method for finding bounds on critical values for non-attractive interacting particle systems. J. Phys. A: Math. Gen. **31**(41), 8323 (1998)

Takeda, K., Tanahashi, M., Unno, H.: Self-repairing mechanism of plastics. Sci. Technol. Adv. Mater. **4**(5), 435–444 (2003)

von Neumann, J., Morgenstern, O.: Game Theory and Economic Behavior. Princeton University Press, Princeton (1944)

von Neumann, J., Burks, A.W.: Theory of Self-reproducing Automata (1966)

Watts, D.J.: Six Degrees: The Science of a Connected Age. WW Norton & Company, New York (2004)

Watts, D.J.: Small Worlds: The Dynamics of Networks Between Order and Randomness. Princeton University Press, Princeton (1999)

Chapter 2
Incentives for Repair in Self-Repair Networks

Abstract This chapter discusses when selfish agents begin to cooperate instead of defect, focusing on a specific task of self-maintenance. To consider the incentive for repair in a game theoretic framework, the Prisoner's Dilemma is introduced in a two-nodes model for the network cleaning problem where a collection of agents capable of repairing other agents by modifying their contents can clean the collection. With this problem, cooperation corresponds to repairing other agents and defect to not repairing. Although both agents defecting is a Nash equilibrium—no agent is willing to repair others when only the repair cost is involved in the payoff —agents may cooperate with each other when system reliability is also incorporated in the payoff and with certain conditions satisfied. The incentive for cooperation will be stronger when a system-wide criterion such as availability is incorporated in the payoff.

Keywords Reliability engineering · Game theory · Mechanism design · Nash equilibrium · Prisoner's dilemma · Hamilton's rule · Kin selection · Multi-agent systems · Mutual repair · Autonomous distributed systems

2.1 Introduction

If von Neumann had worked on introducing active elements (assuming repairing capability) in his research on biological robustness (e.g., probabilistic logic), reliability theory would be more tailored for recent artificial systems involving networked machines. But he left a fundamental framework for active agents, namely game theory. The first step toward a self-repair network is assumed to be a

---

Most of the results of this chapter are presented in Ishida (2007).

© Springer International Publishing Switzerland 2015
Y. Ishida, *Self-Repair Networks*, Intelligent Systems Reference Library 101,
DOI 10.1007/978-3-319-26447-9_2

capability of repair other than classical assumptions for each agent subject to failure and hence passive elements of being repaired. Thus, the fundamental difference from the conventional reliability theory is the assumption of the active aspect in addition to the passive aspect in the nodes. This can be captured as an extension of reliability theory and at the same time as a specialization (for networked computer systems) of the theory. This is made possible by modeling self-involvement of the self-action models (Chap. 1) as follows:

- Self-involvement;
- Autonomous and distributed systems with selfish agents;
- Asymmetry of existence and non-existence.

The model in this chapter is related to asymmetry of existence and non-existence, for it deals with availability and reliability, which are concepts reflected from the real existence space to the functional space. Indeed, the concept of availability is a matter of survival not only for each machine but also for a cooperative collective of machines. In order to consider selfish agents, we need to confirm incentives for the selfish agents to seek. We use the word "agent" when we need to note that the entity is autonomous and hence capable of actions such as repairing and capable of becoming selfish. We also use the word "node" when we need to consider the network structure.

For the self-repair networks, the first question is: even if a framework of self-repair is available, are there any nodes (computers) which would repair other nodes by sacrificing their own resources? Thus, the problem of this chapter is:

> Are there any incentives for a node of the self-repair network to repair other nodes by sacrificing their own resources?

This chapter explores possible incentives by extending the interest of the self-node in space and time. A hint can be gained from the theory of altruism found in social insects (Hamilton 1963). Hamilton noted: "The theory of kin selection defines how an individual values the reproduction of a relative compared with its own reproduction (Hamilton 1964)." With regard to self-repair networks, the remark can be interpreted as: how each node values the assignment of its resources to related nodes compared with its own use. This may be a matter of "exchange rate" as acutely pointed out in an economic theoretical grounding of social evolution by Frank (Frank 1998). We will revisit an evolutionary framework in Chap. 4 involving strategies but this chapter concentrates on incentives for repairing.

Technically, we intend to extend measures of reliability and availability in reliability engineering [e.g., (Shooman 1968; Barlow and Proschan 1975; Anderson and Randell 1979)] so that those of mutually cooperative (repair and being repaired) machines may be measured.

Unexpected growth of large-scale information systems such as the Internet suggests that an open and evolutionary environment for selfish agents will lead to collective phenomena. The Internet is undoubtedly one of the most complex and large-scale artifacts that humans have ever invented. An examination on how the Internet has been built and grown suggests that systems of this complexity may be built not by a usual design but by its own growing logic that not even the designer foresaw before its maturation: a synthetic view that a self-repair network could be embedded in the field.

Since the Internet has formed itself as a field that allows many selfish activities, several utilities and protocols have converged on what may be called the "Nash equilibrium" from which no players want to deviate (Nash 1951, 1953; Nash 1950a, b). In Papadimitriou (2001), a problem for the network protocol is explained, which will lead to economic models that allow the current Internet to exist as an equilibrium point because of its simplicity in permitting distributed and free joining to the network. These studies shed new light on computational intelligence. That is, rather than implementing an intelligent program, design a field in the Internet that allows intelligent systems to emerge as the Nash equilibrium of the Internet field.

Further, the game theoretic approaches to the Internet reveal that obtaining some Nash equilibrium is computationally hard. This fact, looked at from the opposite side, would indicate that a computationally difficult task may be solved by selfish agents. Resource allocation, for example, which is computationally tough, may be solved by a market mechanism in which many selfish agents participate. Mechanism Design, a subfield of economics, has been studied (Hurwicz and Reiter 2006; Maskin 2008; Myerson 1988, 2008) and has been extended to Algorithmic Mechanism Design (Hershberger and Suri 2001; Nisan and Ronen 1999) and to Distributed Algorithmic Mechanism Design (Feigenbaum et al. 2001; Feigenbaum and Shenker 2002; Feigenbaum et al. 2002).

This chapter makes an initial attempt at embedding a computational intelligence in the Internet field by selfish agents; that is, whether selfish agents can ever cooperate and even converge on some tasks. Selfish routing and task allocation have been studied extensively in the computational game community, but can agents ever take care of themselves in the first place? We first pose the problem of self-maintenance in an agent population, and then a game theoretic approach will be tested to determine whether or not cooperation would occur or under what conditions cooperation would occur.

While this chapter amounts to a microscopic analysis focusing on conditions when two interacting agents have an incentive to cooperate (i.e. mutually repair), Chap. 4 amounts to a macroscopic study on a network with many interacting agents.

Section 2.2 discusses the motivations and a paradigm of the present research, and describes the problem of cleaning a self-repair network. Section 2.3 discusses the incentives for selfish agents to cooperate based on system reliability and availability of mutually repairing agents that do not have recognition capability. Section 2.4 discusses when and how the selfish agents will cooperate based on the result of Sect. 2.3.

2.2   Economic Theory for Selfish Agents

The game theoretic approach has demonstrated its power in the field of economics
and biology. The Internet has already reached a level of complexity comparable to
economic systems and biological systems. Moreover, an agent approach permits a
structural similarity where selfish individuals (in the economic system of the free
market) and selfish genes (in biological systems) cooperate or defect in an open
network where many things have been left undetermined before the convergence.

Economic approaches have been actively studied in the distributed artificial
intelligence community [e.g. (Boutilier et al. 1997; Walsh and Wellman 1998)], and
their application to auction may be a successful domain [e.g. (Parkes and Ungar
2000)]. Economic approaches, and a game theoretic approach in particular, have
been extensively studied in the algorithm and computation community and are
having an impact on network applications. Rigorous arguments with the equilibrium concepts, the Nash equilibrium among others, are building a basic theory for
economic aspects of the Internet. The cost of selfish routing has been estimated by
using the extent to which the selfish routing might be degraded at the equilibrium
(Nash equilibrium from which no one wants to deviate) relative to the optimal
solution, as imagined from the traffic congestion caused by most cars want to use
the one shortest path. Protocols such as TCP (Akella et al. 2002), Aloha, CDMA
and CSMA/CA have been studied. Packet forwarding strategies in wireless Ad Hoc
Networks can also be recast in the framework. Network intrusion detection has also
been investigated (Kodialam and Lakshman 2003) in the framework of a
two-players game: Intruder and Defender.

What has been computed by a market mechanism or more generally by a collection of selfish agents turned out to be hard to obtain by computation (as a typical
example: prices of commodities as an index for resource allocation). This fact
indicates that the market economy, or more generally free and hence selfish agents
properly networked, has the potential for computing something that could be hard
when approached otherwise. Also, the fact that the eradication of the planned
economy by the market economy and that the market economy remains in spite of
perturbations suggests that the market economy may be "evolutionarily stable"
(Maynard Smith 1982) within these economic systems.

This fact further indicates that a problem-solving framework by properly networked selfish agents may have some advantage over other usual problem-solving
frameworks such as those organized by a central authority. Also, solutions can be
obtained almost for free or as a byproduct of the problem solving mechanism, or
solutions are almost inseparably embedded in the solving mechanism. The above
two observations encourage the recasting of problems which have been known to
be computationally hard or problems difficult to even define properly and approach,
such as attaining self-repair systems.

Studies with agents usually assume that agents can be autonomous, hence
allowing different rules of interactions: heterogeneous agents. We further assume
that agents are selfish in the sense that they will try to maximize the payoff for the

agent itself. Thus, agents can be broader than the program or software and they involve users that are committed to the agents. Agents may include not only programs but also humans (end-point users and providers running autonomous systems for the Internet) behind the programs. Mutually supporting collectives may emerge as a result of the interplay among agents. Spam mail, computer viruses and worms may be called (malicious) agents, but they are not mutually supporting collectives; they are rather parasitic lone wolves. However, DDoS (Distributed Denial of Service) attacks and some distributed viruses and worms, however, can be considered collectives.

The idea developed here can apply not only to the Internet but also to other information networks such as sensor networks, as long as they can be placed in the model.

The models presented in this chapter have the following components:

M1. States: Agents have two states (0 for normal; 1 for abnormal). The state will be determined by the action and state of interacting agents.
M2. Actions: Agents have two actions (C for cooperation; D for defection).
M3. Network: Agents (nodes) are networked and agents can act only on the connected (and directed by arcs) agents (neighbor nodes).

Actions may be controlled uniformly or may be determined by the acting agent itself in a selfish agent framework so that the payoff assigned to each agent will be maximized. The network may be defined (and visualized as well) explicitly with a graph or implicitly by specifying the neighbor agents (e.g., the lattice structure as in cellular automata and the dynamical network as in scale-free networks).

The network cleaning problem considered here assumes a self-repair network composed of nodes capable of repairing other nodes by modifying the state of the target node (such as resetting, overwriting memory content or even the possibility of re-programming as long as it can be done through the network). Since agents throughout this book are assumed not to have recognition capability, source nodes (repairing agents) can be abnormal and target nodes (agents being repaired) can be normal. Hence mutual repairing without recognition could cause spreading rather than eradication of abnormal states.

Since we focus on the self-maintenance task by mutual repair, cooperation and defection correspond to repairing and not repairing, respectively.

In the agent based approach, we place the following restrictions similar to immunity-based systems (Ishida 2004):

• Local information: For each immune cell mounting a receptor or a receptor itself (antibody), only matching or not (some quantitative information on degree of matching is allowed) can be provided as information.
• No a priori labeling: For an immune cell or antibody, an antigen is labeled neither as "antigen" nor as "nonself."

Because of these two restrictions, we face the "double-edged sword" in this chapter (and throughout the book), since the effectors part (repairing by copying) could harm rather than cure based on local information. This double-edged sword problem (Ishida 2005) (Chap. 3) may be more significant than that the self-recognition model of immunity-based systems because we do not assume recognition capability (that could avoid adverse effects) here as assumed in immunity-based systems. Actions of agents are motivated by selfishness (payoff) rather than the state of the target.

In the following sections, we use a Markov model used for reliability theory as a microscopic model that incorporates M1, M2 and M3 above. The microscopic model focuses on the incentive for cooperation while keeping the network simple with only two interacting agents.

2.3 A Microscopic Model: Negotiation Between Agents

The microscopic model of self-repair networks is based on the concepts of reliability theory such as fault probability, reliability, system reliability and availability. The model also uses a game theoretical framework to consider the network cleaning problem raised in Chap. 1. The model assumptions are as follows:

- Fault: Each node becomes abnormal independently and randomly at a constant rate in a unit time.
- Repair: Each node will repair other nodes at a constant rate in a unit time. Repairing involves consumption of resources of the repairing node. Only normal nodes can repair successfully. Abnormal nodes can also repair successfully but at a constant rate smaller than one.

We need the following game theoretic concepts before defining the microscopic model (two-nodes model).

2.3.1 Prisoner's Dilemma

In solving the problem of cleaning the contaminated network by mutual copying, another problem (other than the double-edged sword) is that each autonomous (and hence selfish) node may not repair others and thus fall into a deadlock waiting for other nodes to repair. This situation is similar to that of the Prisoner's Dilemma that has been well studied in game theory and has been applied to many fields.

The Prisoner's Dilemma (PD) is a game played just once by two agents with two actions (cooperation, C, or defect, D). Each agent receives a payoff R, T, S, P (Table 2.1) where $T > R > P > S$ and $2R > T + S$. Because of the inequality $T > R > P > S$, each player will take action D no matter what action the adversary takes, for the player will get the higher payoff. Since the situation is symmetrical for

Table 2.1 The payoff matrix of the Prisoner's Dilemma (R, S, T, P are payoffs to agent 1)

Agent 1	Agent 2	
	C	D
C	R (Reward)	S (Sucker)
D	T (Temptation)	P (Punishment)

both players, they will take action D, resulting in payoff P which is lower than payoff R when both players cooperate. The inequality $2R > T + S$ prevents one player and another from taking actions C and D respectively (and dividing the payoff equally afterward) whose averaging payoff is $(T + S)/2$.

In the Iterated Prisoner's Dilemma (IPD) (Axelrod 1987, 1984), each iterated action is evaluated many times. In the spatial Prisoner's Dilemma (SPD) (Nowak and May 1992) (Chap. 4), each site in a two-dimensional lattice corresponding to an agent plays PD with its neighbors, and changes its action depending on the total score it received.

When the above inequalities are satisfied, the case where both players take action D is a Nash equilibrium from which neither player wants to deviate. In our model, no agent wants to repair other agents. When trapped in this Nash equilibrium, all agents remain silent, and hence all the agents will eventually enter the abnormal state. With this state of all agents abnormal, there will be no hope of recovering. Incorporation of a system theoretic framework will reveal not only the Nash equilibrium with all agents taking D actions, but also the absorbing state with all agents abnormal from which no recovery can happen.

Other than the prisoner's dilemma with the structure of a payoff matrix satisfying $T > R > P > S$ and $2R > T + S$, other structures such as the Hawk-Dove game (Smith and Price 1973) have been discussed. We mainly focused on PD, for it often applies to society and even to information systems where selfish agents mainly seek their own interest. However, adopting other game structures such as Hawk-Dove and the public goods game (agents not willing to repair other agents may be called free-riders) will be interesting challenges.

2.3.2 Models of System Reliability by Birth-Death Process

Models of system reliability consider how the component reliability and the system structure affect the system reliability. Reliability is an essential probabilistic concept (which is complementary to fault probability) in reliability theory. Reliability of the system can be defined as the probability of being normal (not being faulty or abnormal) at a snapshot, or as the rate of being fault-free during a unit time as a probabilistic process. For a repairable system, yet another probabilistic concept of availability is important, which is related not only with the fault rate but also with the repair rate. Probability (probabilistic measure) is not directly related to time and is meant to indicate the tendency of an event to occur relative to other events (hence, it has no dimension). Rate in a probabilistic process (a Markov process, the birth-death process in particular), on the other hand, is related to time, and is meant to indicate a tendency of occurrence during a unit of time. Although the probability is normalized as values ranging from 0 to 1, the rate can be larger than 1.

Here we face the intrinsic problem of mapping probability to time again. In reliability engineering, the mapping is carried out based on an experimental knowledge of statistical data such as how often a component of interest failed during a specific time interval. In this chapter, we mainly use the rate instead of the probability.

Using conventional notations in reliability theory, λ and μ indicate the failure rate (rate of becoming abnormal) and the repair rate respectively. For example, if the system has only one component which fails with the rate λ, the transition rate from normal state to abnormal state during time Δt is $\lambda \Delta t$ (*death*). Likewise, the state transition rate from abnormal state to normal state is $\mu \Delta t$ (*birth*). Thus, this simplest model with one component being abnormal as well as being normal (repaired) has the following state transition matrix and state transition diagram (Fig. 2.1).

$$\begin{pmatrix} 1 - \lambda \Delta t & \lambda \Delta t \\ \mu \Delta t & 1 - \mu \Delta t \end{pmatrix}$$

Let us consider a Markov model with the continuous time variable. Letting $p_i(t)$ be a probability of being a state i ($i = 0$ for normal and 1 abnormal) at a continuous time t, the following equations describe the state transition:

$$p_0(t + \Delta t) = (1 - \lambda \Delta t)p_0(t) + \mu \Delta t\, p_1(t) + \mathrm{O}(\Delta t)$$
$$p_1(t + \Delta t) = \lambda \Delta t p_0(t) + (1 - \mu \Delta t)p_1(t) + \mathrm{O}(\Delta t)$$

where $\mathrm{O}(\Delta t)$ denotes all the terms with second order or higher of Δt.

In the limit of Δt converging on 0 (denoted by $\Delta t \to 0$), the above equations may be written as a differential (Kolmogorov) equation:

$$d\boldsymbol{P}(\mathrm{t})/\mathrm{dt} = \boldsymbol{P}(\mathrm{t})\boldsymbol{M}^t$$

where the time dependent vector variable

$$\boldsymbol{P}(\mathrm{t}) = (p_0(t), p_1(t)),$$

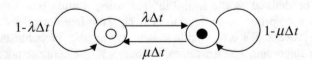

Fig. 2.1 State-transition diagram for one component to be abnormal (faulty) and to be normal (repaired). The white circle indicates a normal node and the black one an abnormal node

and M^t is the transpose of the following matrix M:

$$M = \begin{pmatrix} -\lambda & \mu \\ \lambda & -\mu \end{pmatrix}.$$

Stationary distribution $P(\infty) = (p_0(\infty), p_1(\infty))$ can be obtained by making $d\mathbf{P}(t)/dt = 0$, hence solving the linear equation $P(\infty) M^t = 0$. In this simplest example of the birth-death process, $P(\infty) = (\mu/(\lambda + \mu), \lambda/(\lambda + \mu))$, however the stationary distribution itself simply follows from the symmetry of the model (death and birth is just a matter of labeling the different symbols λ and μ, and the exchange symmetry holds).

2.3.2.1 Mutual Repairing with Selfish Agents

Consider a model with only two agents i ($i = 1, 2$) that are capable of repairing the other agent. Using conventional notations again in reliability theory, λ and μ indicate the failure rate (rate of becoming abnormal) and the repair rate respectively. In considering two agents, a repair action must be considered as an interaction from the repairing agent to the agent being repaired, while the failure event (of normal agents becoming abnormal) occurs within an agent. The double-edged sword framework allows agents that are capable of repairing other agents, but when the repairing agents are themselves abnormal they will cause the target agents to be abnormal (spread contamination) rather than repairing. Thus the state-transition diagram as a Markov model is as shown in Fig. 2.2. Let μ_i denote the repair rate done by agent i, and let α (<1) indicate the repair success rate when repair is done by an abnormal agent. This repair success rate is in fact a probability, for it switches the successful repairs with the rate $\alpha \mu_i$ and the failed repair with the rate $(1-\alpha)\mu_i$. Repairs by normal agents are assumed to be always successful. For simplicity, both a failure event and a repair action do not occur simultaneously. The corresponding Kolmogorov equation is:

$$\frac{d\mathbf{P}(t)}{dt} = \mathbf{M}\mathbf{P}(t)$$

where the time dependent vector variable

$$\mathbf{P}(t) = (p_{00}(t), p_{01}(t), p_{10}(t), p_{11}(t))^T$$

comprises a component $p_{s_1 s_2}(t)$ denoting a probability of agent 1 being S_1 and agent 2 being S_2 at time t where $S_1, S_2 \in \{0, 1\}$ (0: normal; 1: abnormal). $\mathbf{M}$ is a transition matrix corresponding to the state-transition diagram shown in Fig. 2.2.

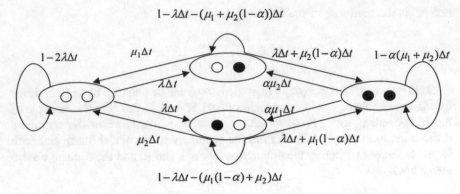

Fig. 2.2 State-transition diagram for the mutual repairing two agents system. White circles indicate normal nodes and black ones abnormal nodes

$$
M = \begin{pmatrix}
-2\lambda & \mu_1 & \mu_2 & 0 \\
\lambda & -\lambda - (\mu_1 + (1-\alpha)\mu_2) & 0 & \alpha\mu_2 \\
\lambda & 0 & -\lambda - ((1-\alpha)\mu_1 + \mu_2) & \alpha\mu_1 \\
0 & \lambda + (1-\alpha)\mu_2 & \lambda + (1-\alpha)\mu_1 & -\alpha(\mu_1 + \mu_2)
\end{pmatrix}
$$

For a game theoretic argument, it is further assumed that an agent must decide whether it will repair others or not, corresponding to cooperation and defection in the Prisoner's Dilemma. For agent i, $C_i = 1$ if it repairs another agent, and 0 otherwise. Let $P_i(C_1, C_2)$ denote a probability of agent i being normal when agent i's action is C_i. A simple calculation yields the steady-state probability of $P_i(C_1, C_2)$ as listed in Table 2.2.

When abnormal agents are assumed to do nothing and remain silent as in the case of mechanical systems, then both agents in the abnormal state is the absorbing state, and hence the steady-state probabilities of $P_i(C_1, C_2)$ are all 0. In this model, all agents will be abnormal eventually no matter whether cooperation takes place or not. Thus, we assume that even abnormal agents may repair when they take action C. When one agent repairs another agent, the repairing rate is assumed to be the same one: μ. That is, if both agents cooperate (repair), then $\mu_1 = \mu_2 = \mu$. If agent 1 cooperates, but agent 2 does not, then $\mu_1 = \mu$ but $\mu_2 = 0$.

Table 2.2 Steady-state reliability of each agent when mutual repairing is involved $\rho = \frac{\lambda}{\mu}$

	$C_2 = 1$	$C_2 = 0$
$C_1 = 1$	$P_1(1, 1) = \frac{\alpha}{\rho + \alpha}$	$P_1(1, 0) = 0$
	$P_2(1, 1) = P_1(1, 1)$	$P_2(1, 0) = P_1(0, 1)$
$C_1 = 0$	$P_1(0, 1) = \frac{\alpha}{\rho + 1}$	$P_1(0, 0) = P_2(0, 0) = 0$
	$P_2(0, 1) = P_1(1, 0)$	

Table 2.2 can be regarded as a payoff matrix of the two-players game. If we simply regard $P_i(C_1, C_2)$ as agent i's payoff when actions C_1, C_2 are taken, mutual repairing may happen because of the inequalities:

$$P_1(1, 1) > P_1(0, 1) > P_1(1, 0) = P_1(0, 0),$$
$$P_2(1, 1) > P_2(1, 0) > P_2(0, 1) = P_2(0, 0).$$

While the action does not make any difference (e.g. for the agent 1, $P_1(1, 0) = P_1(0, 0)$) when another agent does not cooperate, the agent should certainly cooperate when another agent cooperates (e.g. for agent 1, $P_1(1, 1) > P_1(0, 1)$). This is because by raising the reliability of others, the repairing by them to the self becomes more effective, a circular effect.

Let us take the cost of repairing into consideration. Although both D is a Nash equilibrium when there is a positive repair cost, the agents do not have an incentive to remain in the both D when the repair cost is negligible.

Let us focus on the payoff. Then agent 1, for example, will choose its action C_1 to maximize:

$$P_1(C_1, C_2) - c \cdot C_1,$$

where c is a cost of repairing relative to the benefit measured by the reliability of itself. Incorporating a cost for cooperation would naturally bias the situation toward more defect-benefiting. When the adversary agent defects, an agent simply loses the cost for cooperation if it cooperates. However, there is still a chance for mutual cooperation when the opponent cooperates: $P_1(1, 1) - c > P_1(0, 1)$ holds when the cost relative to benefit satisfies:

$$\frac{\alpha(1 - \alpha)}{(\rho + 1)(\rho + \alpha)} > C.$$

Selfishness of an agent is reflected on the objective function that the agent will maximize, and the reflection is not a trivial task. The above agents are shortsighted in implementing the selfishness. Foresighted agents would consider the event of another agent's failure as losing the chance of being repaired by the agent, and the extinction of all normal agents as a fatal event that should be avoided by paying a high cost. If the repairing by abnormal agents does not happen, extinction of normal agents is an absorbing state from which no other normal state will arise when the repair success rate by abnormal agents α is close to 0 (repairs by abnormal agents do not virtually succeed).

Figure 2.3 plots the difference $P_1(1, 1) - P_1(0, 1)$ when the repair success rate by abnormal agent α changes from 0 to 1 and $\lambda = 10^{-4}$, $\mu = 10^2 \lambda$ are fixed. There is a strong incentive for agent 1 to cooperate when the success rate α is about 0.1. The incentive decreases almost linearly when the rate exceeds 0.2 in this case, which indicates that reliable repairs by abnormal agents would not promote cooperation.

Fig. 2.3 Plot of the
difference when the repair
success rate by abnormal
agents α changes from 0 to 1
and $\lambda = 10^{-4}$, $\mu = 10^2 \lambda$ are
fixed

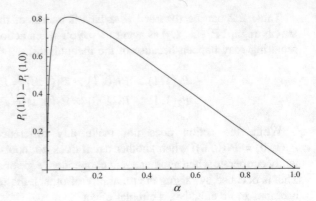

Table 2.3 Steady-state availability AV where availability is a probability that at least one agent is normal

	$C_2 = 1$	$C_2 = 0$
$C_1 = 1$	$AV(1, 1) = \frac{\alpha(2\rho+1)}{(\rho+1)(\rho+\alpha)}$	$AV(1, 0) = \frac{\alpha}{\rho+1}$
$C_1 = 0$	$AV(0, 1) = \frac{\alpha}{\rho+1}$	$AV(0, 0) = 0$

Let us consider the availability (the probability that the system is available at the time, hence in our model, the probability that at least one agent remains normal). Let $AV(C_1, C_2)$ denote the availability when agent i's action is C_i. Technically, we use *limiting average availability* (e.g., (Barlow and Proschan 1975)) as a payoff for each agent, then there will be a stronger incentive to cooperate when the other agents cooperate, since the difference $AV(1, 1) - AV(0, 1)$ is larger than the difference $P_1(1, 1) - P_1(0, 1)$ as shown in Table 2.3.

This indicates that even selfish agents will be more likely to cooperate if they take a systemic payoff that evaluates cost and benefit in a more system-wide fashion and a longer-term basis: the beginning of self-organization of mutually supporting collectives.

2.4 Discussion

2.4.1 Nash Equilibrium

The worst-case analysis (Koutsoupias and Papadimitriou 1999) uses a Nash equilibrium as a solution when tasks are left to selfish agents. The cost for the Nash equilibrium relative to the optimized solution has been proposed to measure the cost of "anarchy" (Koutsoupias and Papadimitriou 1999). This chapter rather focused on the self-maintenance task, self-repairs by mutual copying in particular, and discussed when selfish agents begin to cooperate. Further discussions are needed on when these selfish agents organize themselves into mutually supporting collectives.

The present research has two significances: one engineering and another theoretical. For engineering, computing paradigms such as distributed computing systems (Farber and Larson 1970), grid computing (Foster and Kesselman 2001; Foster and Kesselman 2003; Foster et al. 1998) and parasitic computing (Barabási et al. 2001) provide a background. When grid computing becomes dominant for large-scale computing, what we call agents (autonomous programs that can move from nodes to nodes) will become like processes in the Unix operating system. One important difference is that agents may be selfish, and will not be organized with a central authority as is done in conventional operating systems. Then, the organization of selfish agents will become an organization with a weakest central authority, or even with a distributed authority as seen in the free market economy. Naturally, information processing with selfish agents will be imperative, thus making the game theoretic approach and economic approach such as selfish task allocation and routing important.

Another significance is that it will provide an organizational approach to artificial life (a life-like form which has some identity hence boundary). Self-organization of selfish agents will be more than a mere collection of independent agents, but rather a cluster of cooperative agents. This would reveal an intrinsic logic and process that selfish agents form multi-agent organisms, similarly to multi-cellular organisms. The game theoretic approach will provide a threshold and a mechanism for selfish agents to develop into cooperative agents when payoffs are recast in a broader context of time and space.

2.4.2 Hamilton Rule as a Condition for Altruism

In evolutionary biology, many theories have been proposed that explain the altruistic behaviors of individuals. One of them is kin selection where altruistic behaviors among relatives can be explained by extending fitness to inclusive fitness with relatedness. Hamilton's rule (Hamilton 1964) is formulated as follows:

$$rB > C$$

where the relatedness r (the kin selection coefficient of relatedness between altruistic agent and recipient agent) can be measured by genetic distance. B is the reproductive benefit to the recipient by the altruistic behavior and C is the cost for the altruistic behavior. Frank (Frank 1998) has applied Hamilton's rule to social evolution by extending this relatedness r to a measure generalized from the genetic distance. The rule can explain the extraordinary sex ratio observed in social insects such as honey bees (Hamilton 1963).

From a cost-benefit point of view, Hamilton's rule can be a simple condition for an action of an agent where the action will be carried out when the benefit exceeds the cost. Let us borrow Frank's understanding of relatedness in the framework of economic optimization through *exchange rate* (Frank 1998). One notable point is that the benefit is not directly oriented toward the self but indirectly received, hence the benefit must be discounted by multiplying by a discount rate r. Thus, the relatedness can be regarded as a spatial version of discount rate (usually discount rate is related to time; the value at a future time is discounted compared to the value at the present time). In the model of this chapter, the indirect benefit (discounted by relatedness r) from the interacting agent is measured by the difference of reliability: $P_1(1, C_2) - P_1(0, C_2)$ or the difference of availability: $AV_1(1, C_2) - AV_1(0, C_2)$. One challenge would be to compare the benefit by actions toward itself and actions toward a neighbor, rather than comparing the benefit by actions toward the neighbor with that of doing nothing.

The model proposed in this chapter recast a possible mechanism to promote altruistic behavior among nodes in a network based on Hamilton's rule where the relatedness r can be measured by a distance in the network, that is, how close the nodes are in the network (how direct the exchange of resources can be). In the cost-benefit analysis of the previous Sect. 2.3, the relatedness $r = 1$ of Hamilton's rule when the node is in the neighbor (directed by an arc) and $r = 0$ otherwise. The condition for mutual repair is also the cost-benefit condition for the action of repairing the neighbor nodes where the benefit is measured by the increase of the reliability (or availability) discounting the fact that the repairing effect is not directly to the self but indirectly through the neighbors. Since the self-repair network uses a network to express the structure, the benefit of being repaired is discounted if the repaired node is far from the repairing node. But why would the repairing node (the self-node) not use the entire resource for repairing itself (the self-node)? We suppose the following rationale for the diversification of the risk specific to the self-repair network: when the node repairs itself there are only two cases in the repair pattern, i.e. a normal node repairs the normal node; or an abnormal node repairs the abnormal node. When the node repairs the neighbor nodes, even though the benefit is indirect (and hence discounted), there are two other cases in the repair pattern, i.e., a normal node repairs an abnormal node; or an abnormal node repairs a normal node. The former is an *edge* (advantage) for a custom repair of the double-edged sword and the latter is another *edge* (disadvantage) that is inevitably associated with mutual repair in the self-repair network without recognition. We will compare mutual repair and self-repair in the self-repair network in Chap. 9.

Hamilton's rule (the condition for altruism) may be viewed from another way, that is, the self can be extended to a system connected by the mutual repair: *quasi-self*. We will consider re-modification of the payoff called *systemic payoff* in Chap. 5, which amounts to considering the *availability* (as a system) rather than the *reliability* (of a node) in the theory of reliability.

Although this chapter focused on the incentive for a node to repair other nodes, the mechanism of spreading the repairing *trait* will be considered in Chap. 4 (direct mechanism where repairing is inevitably associated with copying of the repairing strategy) and in Chap. 5 (indirect mechanism where the node copies the strategy of the neighbor node who earned the largest payoff).

2.5 Conclusion

For large-scale information systems, a game theoretic approach is important, since it will give results concerning what would happen when selfish agents are involved. However, what is "selfish" depends on the context and the environment. This research assumes that selfish agents try to maximize their payoff. Then the next problem is to set the payoff function reflecting the context and the environment. We discussed the cases when only a repair cost, the system reliability, and a more systemic evaluation such as the availability (*limiting average availability*) are incorporated. Incentives to cooperation increase when a more systemic evaluation is involved in the payoff. Specifically, if an agent sticks to a short-sighted payoff such as the repair cost, the agent will lose the partner that would repair the agent when it becomes abnormal, or even worse, all the agents will eventually become abnormal and will forever lose the chance of being repaired.

The current research should be further developed through studies on how and when mutually supporting collectives emerge in large-scale information systems such as the Internet.

It is important to note that the models not only in this chapter but throughout this book have limitations in directly applying them to real situations due to the fact that they sacrifice reality for simplicity in order to focus on the problem in question. For example, the model parameters are expressed as constant values, however, they could change or adapt to the environment, or they may even be difficult to be expressed as parameters. For the model in this chapter, the key parameter of repair success rate and other parameters such as repair rate, could change over time. But the essence of the self-repair network as a model resides in the asymmetry (of existence and non-existence) that the repair success rate by abnormal agents is less than that by normal agents.

While this chapter views the self-repair network from the nodes within it and attributes incentives for cooperation (repair) to the relatedness of nodes reminiscent of Hamilton's theory of altruism, one can also view the self-repair network from outside and attribute incentives for cooperation to the fact that the network is so integrated that cooperation is the selfish act of helping the network itself.

Although we have shown that there is an incentive for a node of the self-repair network to cooperate (repair other nodes) in terms of reliability engineering, whether mutual repair can be realized or not is another story. We will further study game theoretically with the spatial prisoner's dilemma, and examine how and when the strategy with cooperation remains or should remain, in Chap. 4. Before that, we will investigate the network cleaning problem with mutual and non-strategic repair (meaning uniform repairing carried out independently from neighbors' actions) in the following Chap. 3.

References

Akella, A., Seshan, S., Karp, R., Shenker, S.: Selfish behavior and stability of the Internet: a game-theoretic analysis of TCP. Comput. Commun. Rev. **32**(4), 117–130 (2002)

Anderson, T., Randell, B.: Computing Systems Reliability. CUP Archive (1979)

Axelrod, R.: The Evolution of Cooperation. Basic Books, New York, NY (1984)

Axelrod, R.: The evolution of strategies in the iterated prisoner's dilemma. Dyn. Norms, 199–220 (1987)

Barabási, A.L., Freeh, V.W., Jeong, H.W., Brockman, J.B.: Parasitic computing. Nature **412** (6850), 894–897 (2001). doi:10.1038/35091039

Barlow, R.E., Proschan, F.: Statistical theory of reliability and life testing: probability models. I: DTIC Document (1975)

Boutilier, C., Shoham, Y., Wellman, M.P.: Economic principles of multi-agent systems. Artif. Intell. **94**(1–2), 1–6 (1997)

Farber, D.J., Larson, K.: The architecture of a distributed computer system-An informal description, Technical Report. University of California, Irvine, CA (11) (1970)

Feigenbaum, J., Shenker, S.: Distributed algorithmic mechanism design: Recent results and future directions. September (2002)

Feigenbaum J., Papadimitriou C., Sami R: A bgp-based mechanism for lowest-cost routing. In: 21st ACM Symposium on Principles of Distributed Computing, pp. 173–182. ACM Press, Monterey, CA (2002)

Feigenbaum, J., Papadimitriou, C.H., Shenker, S.: Sharing the cost of multicast transmissions. J. Comput. Syst. Sci. **63**(1), 21–41 (2001). doi:10.1006/Jcss.2001.1754

Foster, I., Kesselman, C.: Computational grids—Invited talk. Lect. Notes Comput. Sci. **1981**, 3–37 (2001). (Reprinted from The Grid: Blueprint for a new computing infrastructure, 1998)

Foster, I., Kesselman, C., Tsudik, G., Tuecke, S.: A security architecture for computational grids. In: Proceedings of the 5th ACM Conference on Computer and Communications Security, pp. 83–92. ACM (1998)

Foster, I., Kesselman, C.: The Grid 2: Blueprint for a new computing infrastructure. Morgan Kaufmann, (2003)

Frank, S.A.: Foundations of Social Evolution. Princeton University Press, Princeton (1998)

Hamilton, W.D.: The evolution of altruistic behavior. Am. Nat. **97**(896), 354–356 (1963)

Hamilton, W.D.: The genetical evolution of social behaviour. I. J. Theor. Biol. **7**(1), 1–16 (1964)

Hershberger, J., Suri, S.: Vickrey prices and shortest paths: What is an edge worth? In: Proceedings. 42nd IEEE Symposium on 2001 Foundations of Computer Science, pp 252–259, IEEE (2001)

Hurwicz, L., Reiter, S.: Designing Economic Mechanisms. Cambridge University Press (2006)

Ishida, Y.: A critical phenomenon in a self-repair network by mutual copying. In: Knowledge-Based Intelligent Information and Engineering Systems, pp. 86–92. Springer, Berlin (2005)

Ishida, Y.: A game theoretic analysis on incentive for cooperation in a self-repairing network. In: Innovations and Advanced Techniques in Computer and Information Sciences and Engineering, pp. 505–510. Springer, Berlin (2007)

Ishida, Y.: Immunity-Based Systems: A Design Perspective. Springer, New York Incorporated (2004)

Kodialam, M., Lakshman, T.: Detecting network intrusions via sampling: a game theoretic approach. In: INFOCOM 2003. Twenty-Second Annual Joint Conference of the IEEE Computer and Communications. IEEE Societies, pp. 1880–1889. IEEE (2003)

Koutsoupias, E., Papadimitriou, C.: Worst-case equilibria. STACS'99—16th. Ann. Symp. Theor. Aspects Comput. Sci. **1563**, 404–413 (1999)

Maskin, E.S.: Mechanism design: How to implement social goals. Am. Econ. Rev. **98**(3), 567–576 (2008)

Maynard Smith, J.: Evolution and the Theory of Games. Cambridge University Press, Cambridge; New York (1982)

Myerson, R.B.: Mechanism design. Center for Mathematical Studies in Economics and Management Science, Northwestern University, (1988)

Myerson, R.B.: Perspectives on mechanism design in economic theory. Am. Econ. Rev. 586–603 (2008)

Nash, J.: Non-cooperative games. Ann. Math. **54**(2), 286–295 (1951)

Nash, J.: Two-person cooperative games. Econometrica: J. Econometric Soc. 128–140 (1953)

Nash, J.F.: Equilibrium points in n-person games. Proc. Natl. Acad. Sci. **36**(1), 48–49 (1950b)

Nash, J.F.: The bargaining problem. Econometrica: J. Econometric Soc. 155–162 (1950a)

Nisan, N., Ronen, A.: Algorithmic mechanism design. In: Proceedings of the Thirty-First Annual ACM Symposium on Theory of Computing, pp. 129–140. ACM (1999)

Nowak, M.A., May, R.M.: Evolutionary games and spatial chaos. Nature **359**(6398), 826–829 (1992)

Papadimitriou, C.H.: Algorithms, games, and the internet. Automata Lang. Program, Proc. **2076**, 1–3 (2001)

Parkes, D.C., Ungar, L.H.: Iterative combinatorial auctions: Theory and practice. In: Proceedings of the National Conference on Artificial Intelligence 2000, pp. 74–81. Menlo Park, CA; Cambridge, MA; London; AAAI Press; MIT Press; 1999

Shooman, M.L.: Probabilistic Reliability: An Engineering Approach, vol. 968. McGraw-Hill, New York (1968)

Smith, J.M., Price, G.: The logic of animal conflict. Nature **246**, 15 (1973)

Walsh, W.E., Wellman, M.P.: A market protocol for decentralized task allocation. In: Proceedings. International Conference on 1998 Multi Agent Systems, pp. 325–332. IEEE (1998)

Chapter 3
A Phase Transition in Self-Repair Networks: Problems and Definitions

Abstract This chapter reports a critical phenomenon in a self-repair network by mutual copying. Extensive studies have been done on critical phenomena in many fields such as in epidemic theory and in percolation theory in order to identify critical points. However, critical phenomena have hardly been studied from the viewpoint of cleaning up a network by mutual copying. A critical phenomenon has been observed in a self-repair network. Self-repairing by mutual copying is "a double-edged sword" that could cause outbreaks if parameters are inappropriate, and therefore careful investigations are needed.

Keywords Cellular automaton · Critical phenomenon · Percolation theory · Domany-Kinzel model · Network clean-up problem · Uniform repair

3.1 Introduction

Before incorporating the action of repairing in each node, we need to know its macroscopic impact in addition to microscopic incentives (Chap. 2). It is found that involving repair actions at the node level leads to the emergence of a macroscopic critical phenomenon, the double-edged sword. It is indeed double edged, for it could spread abnormal nodes or could wipe out them, and it is indeed "edged," for these two phases are acutely separated by a critical point of the parameter. We use a technique of the probabilistic cellular automaton incorporated in the Domany-Kinzel model, which uses the probabilistic cellular automaton for the Ising model. Another avenue in modeling may be to use the percolation model to identify the condition for the critical point where an infinite cluster emerges.

Critical phenomena in physics such as phase transition can be understood to mean that a significant change in phase takes place (e.g., water evaporates or

---

Most results of this chapter are presented in Ishida (2005).

© Springer International Publishing Switzerland 2015   37
Y. Ishida, *Self-Repair Networks*, Intelligent Systems Reference Library 101,
DOI 10.1007/978-3-319-26447-9_3

freezes) when a parameter (e.g., temperature) crosses a critical point (e.g., boiling point or freezing point). Critical phenomena, however, include not only physical phenomena but also many other phenomena such as the outbreak of epidemics. Percolation theory helps us to understand that state propagation through a medium can be a candidate critical phenomenon. This chapter explores the possibility that the normal state of a computer could percolate through a network by mutual repairing when certain conditions are met.

The essence of a phase transition in physics is a change of state from high symmetric ordered state with high temperature to low symmetric disordered state with low temperature. The ordered parameters play a crucial role in characterizing the symmetry break. Phase transitions fall into two types: the first-order transition involving heat transfer as observed in transitions among gas, liquid and solid, and the second-order transition that does not involve latent heat as observed in ferromagnetic transition and superfluid transition. The term "first-order" comes from the discontinuity of the first-order derivatives of the energy with respect to a parameter (temperature), while the term "second-order" indicates the discontinuity of the second-order derivatives. Critical phenomena include discontinuous behaviors of physical quantities (due to the divergence of correlation length) observed in the neighbor of the critical point for the case of the second-order transition. A fractal structure (or self-similar structure between different scales) is often identified near the critical point. Scaling universality in critical phenomena has been an important issue in creating the phase transition observed in many fields, not only in statistical physics but also in epidemics.

It has been pointed out that the Internet is a scale-free network (Barabási and Frangos 2002), and it has also been suggested that the network is tolerant to random attacks but vulnerable to selective attacks (Dezso and Barabási 2002); hence, selective defense at the hub seems to be effective. Moreover, it has also been pointed out that the critical point (above which the epidemic breaks out) in scale-free networks is 0, thus it is difficult to eradicate computer viruses and worms from the Internet (Dezso and Barabási 2002).

As a countermeasure against selective attacks and the spread of computer viruses when the critical point is 0, selective or even adaptive defense (Chap. 4) can be considered. In physical systems such as mechanical systems, they are repaired by identifying the faulty components and replacing them with non-faulty ones. In information systems, however, they can be repaired by simply copying the clean system to the contaminated system. As a first step toward adaptive defense of information systems, we consider self-repairing of the network by mutual copying.

3.2   Self-Repair by Copying: The Double-Edged Sword

In outbreaks of biological epidemics (Boccara and Cheong 1993; Rhodes and Anderson 1996, 1998), spontaneous recoveries can occur and even give rise to the immune system. However, in information systems such as computer networks, humans must repair the systems or computers must repair each other mutually when contaminated with computer viruses and worms. This chapter deals with the latter case where computers mutually repair each other. However, in this case, computers could contaminate other ones if the repairing computers themselves are contaminated. This is because computers repair software faults by copying, which is quite different from mechanical systems. When a mechanical system suffers a hardware fault, the repairing nodes cannot simply repair others if the repairing nodes themselves are faulty. In contrast, when an information system suffers a software fault, the repairing nodes can repair others simply by copying their content. However, self-repair by copying could spread contamination if the network is already highly contaminated or the infection rate is high. This chapter concentrates on the naive problem of cleaning up the network by mutual copying: when can a contaminated network be cleaned up by mutual copying?

We have studied immunity-based systems and pointed out the possibility of the double-edged sword. For example, a system based on Jerne's idiotypic network framework (Ishida 2004) has recognizing nodes that are also being recognized. Self-nonself recognition involving the self-referential paradox could also lead to the situation of the double-edged sword: recognition done by sufficiently credible nodes is credible, but not credible otherwise.

The repair by copying in information systems is also the double-edged sword and it is an engineering concern to identify when it can really eradicate abnormal elements from the system. This chapter considers cellular automata (CA), probabilistic CA specifically, to model the situation where networked computers mutually repair by copying their content. We will focus on the simplicity rather than the reality of the system.

Studies on the reliability of information systems by a simple model of a probabilistic CA are not new. The results of Gacs (2001) for his probabilistic CA can be summarized as: *"there exists a simple finite-state machine M such that for any computer program π and any desired level of reliability p < 1, the program π can be run with reliability p on a sufficiently large one-dimensional array of copies of M, each one communicating only with its nearest neighbors, even if these machines all have the same small but positive error rate"* (Gray 2001), when networked computers are considered.

The problem considered in this chapter is when can "double-edged" self-repair be used:

If self-repairing could cause abnormal nodes both to decrease and increase, when should repairing be conducted?

To focus on the double-edged sword, we adopt the simplest model assumptions:

- Fault: A node becomes abnormal only when the repair failed.
- Repair: Not only normal nodes but also abnormal nodes can repair with a higher risk of unsuccessful repair. Repair does not require any cost.

- Control: Actions are spatially arranged alternately, repair and being repaired, in a one-dimensional ring network of nodes. The action flips alternately between repair and being repaired step by step for each node. The repair actions are executed on two neighbors simultaneously and independently with a constant rate (repair rate).
- Network: An *N*-node one-dimensional lattice with two neighbors and with the periodic boundary condition.

In other words, the risk of using another "edge" is the only cost for the self-repair.

3.3   Models of Probabilistic Cellular Automata

First, the difference between the microscopic model of Chap. 2 (extended from reliability theory) and the macroscopic model of this chapter (related to models of the theory of interacting particle systems) is examined. As already noted, each node does not become faulty spontaneously by itself (in contrast, the model of Chap. 2 has a positive failure rate); nor is each node infected by other abnormal nodes (the model of Chap. 8 has a positive infection rate). The possibility of becoming abnormal is only by unsuccessful repair, and the repair does not have any cost (the model of Chap. 4 repairs with a positive cost involving resource consumption). Also, the microscopic model of Chap. 2 consists of just two agents, while the number of agents in the macroscopic model exceeds two (finite in computer simulations and possibly infinite in the theoretical model). Because the model from this chapter is a macroscopic model with a huge number of nodes, it has structures (e.g., one-dimensional ring structure) and it has interaction rules (e.g., a set of rules for the networked automaton). Other than structures and interaction rules, synchronicity which determines how the interaction rules are applied to each node is an important component. A mathematically rigorous analysis would be difficult unless already known models or approximation techniques exist; thus we rely heavily on computer simulations. For the computer simulations, we use two types of simulation: an agent-based simulation and a numerical analysis. The latter is possible only when a mathematical model is available, while the former can be done by building a cyber-world based on agents.

In the model, the system consists of nodes capable of repairing other connected nodes. We call the connected nodes "neighbor nodes" based on the terminology of CA. The repairing may be done by copying content to the other nodes, since we are considering application to networked computer systems.

Although mutual repairing and testing may be done in an asynchronous manner, our model considers synchronous interactions for simplicity and for comparison with existing probabilistic CA models (e.g., Bagnoli et al. 1997; Domany and

Kinzel 1984; Vichniac et al. 1986). Each node tries to repair its adjacent nodes, however, since the repairing is done by copying it could make the adjacent nodes abnormal rather than normal.

In a mathematical formulation, the model consists of three elements (U, T, R) where U is a set of nodes, T is a topology connecting the nodes, and R is a set of rules of the interaction among nodes. In this chapter, a set of nodes is a finite set with N nodes, and the topology is restricted to the one-dimensional array as shown in Fig. 3.1 (which could be an n-dimensional array, complete graph, random graph, or even scale-free network) that could have S neighbors for each node with a boundary condition, i.e. the structure of the array is a ring with node 1 adjacent to node N. Also, we restrict the case to each node having a binary state: normal (0) or abnormal (1).

Each node tries to repair the adjacent nodes in a synchronous fashion with a probability Pr (called *repair rate*). As shown in Fig. 3.2, the repairing will be successful with the probability Prn (called *repair success rate by normal node*)

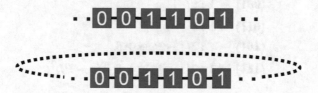

Fig. 3.1 One-dimensional array with two states: normal (*0*) and abnormal (*1*) with an infinite number of nodes (*above*) and with a finite number of nodes with periodic boundary condition (*below*), i.e., a ring structure

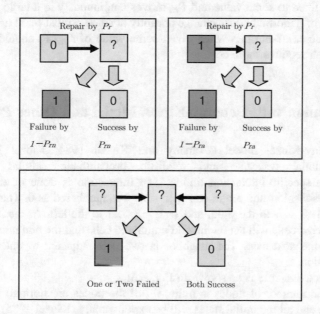

Fig. 3.2 Probabilistic repair by normal nodes (*above, left*) and by abnormal nodes (*above, right*). Repair by two nodes (*below*)

when it is done by a normal node, but with the probability **Pra** (called *repair success rate by abnormal node*) when it is done by an abnormal node (**Pra** < **Prn**) (this is an asymmetry inherited from the existence property). This repair success rate by abnormal nodes reveals another edge of the double-edged sword and plays a crucial role in observing a phase transition here. In this chapter, we assume **Prn** = 1. The repaired nodes will be normal when all the repairing is successful. Thus, when repairing is done by two adjacent nodes, both of these repairs must be successful in order for the repaired node to be normal.

As a probabilistic cellular automaton, the transition rules are as follows (where the self-state is the center in parentheses and the two neighbor states are on the left and right in parentheses. The self-state will be changed to the state indicated to the right of the arrow, with the probability indicated after the colon):

$$(\mathbf{000}) \rightarrow \mathbf{0} : 1,$$
$$(\mathbf{010}) \rightarrow \mathbf{1} : (1 - P_r)^2,$$
$$(\mathbf{001}) \rightarrow \mathbf{1} : \alpha,$$
$$(\mathbf{011}) \rightarrow \mathbf{1} : \alpha + (1 - P_r)^2,$$
$$(\mathbf{101}) \rightarrow \mathbf{1} : 2(1 - P_r)\alpha + \beta,$$
$$(\mathbf{111}) \rightarrow \mathbf{1} : 2(1 - P_r)\alpha + \beta + (1 - P_r)^2,$$

where $\alpha = P_r(1 - P_{ra}), \beta = P_r^2(1 - P_{ra}^2)$.

In such models, it is of interest to determine how the repairing probability P_r should be set when the success probability by abnormal node P_{ra} is given. Also, when P_r is fixed to some value and P_{ra} moves continuously to a large value, does the number of abnormal nodes change abruptly at some critical point or does it just gradually increase? Further, when given some value of P_{ra}, P_r should always be larger, which requires more cost.

3.4   Relation with Domany-Kinzel Model and Other PCA

The Domany-Kinzel model (Domany and Kinzel 1984; Kinzel 1985) is a one-dimensional two-state and totalistic probabilistic cellular automaton (PCA) with specific interaction timing. The interaction is done in an alternated synchronous fashion: the origin cell with state 1 is numbered as 0. The numbering proceeds {1, 2, ...} to the right, and {−1, −2, ...} to the left. At the Nth step the even numbered cells will act on the odd numbered cells and the odd numbered cells will act at the next step. The neighbor is two cells adjacent to oneself without self-interaction.

The interaction rule is as shown in Fig. 3.3:

When the number of nodes is finite, **0** (all the nodes are normal) is only one steady state and all the initial states will be exponentially (Kinzel 1985) attracted to

$(0*0) \rightarrow 0:1, (0*1) \rightarrow 1:p1, (1*1) \rightarrow 1:p2$

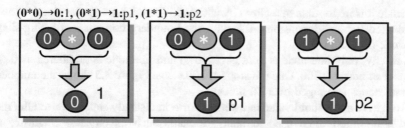

Fig. 3.3 Transition of the Domany-Kinzel model when both sides are *0* (*left*), one side is *1* (*middle*), and both sides are *1* (*right*)

Fig. 3.4 Synchronization of the Domany-Kinzel model. Acting nodes and being acted nodes change alternately in space and time

0 when $p1 < 1$, $p2 < 1$. When the number of nodes is infinite, the system has other steady states with 1 and 0 mixed with a certain ratio, other than all 0 states. Our PCA model can be equated with the DK model when $P_r = 1$ (i.e. nodes always repair) with the parameters: $p1 = \alpha (= (1 - P_{ra}))$, $p2 = \beta (=(1 - P_{ra}^2))$; i.e. the case of the directed bond percolation. For two-dimensional cases, some results of the p-Voter model (Eloranta 1995) are available. Also, if we use the alternated synchronization of the DK-model, the self-repair network has the following synchronization scheme: the repairing node and the being-repaired node are placed alternately in one dimension, and they will alternate as the time step proceeds (Fig. 3.4).

3.5 Agent-Based Simulation and Approximated Results

Viewed as a Markov chain with finite states, the Markov chain assures that the state will eventually converge on the stationary state. Thus, we will obtain the same results for the ratio of normal/abnormal nodes starting from any state if it is not a fixed point. For our model, the state with all normal nodes is a fixed point if $P_{rn} = 1$ and the state with all abnormal nodes is a fixed point if $P_{ra} = 0$. In the following computer simulation, we exclude the latter cases assuming $P_{ra} > 0$.

Since it is impossible to realize CA with infinite nodes, computer simulations are conducted for one-dimensional CA with the boundary condition of a ring-shaped network (Fig. 3.1).

Initially, only one node is normal (state 0) and the node is numbered as 0. The number of nodes is 500. One run stops at 800 steps. Figure 3.5 shows the number of normal nodes (averaged over 10 times).

To derive a threshold value in a simple form in a steady state, we need the mean field assumption often used in many domains such as physics, chemistry and ecology, which assumes that an entity is surrounded by a field whose parameter of interest (e.g., density, concentration, or probability of encountering other individuals) may be assumed constant throughout a space. This assumption is often used to simplify spatial complexity, and to derive a simpler but approximated equation of, for example, a master equation describing a net increase of specific entities. Approximation depending on the assumption is called mean field approximation (MFA) which approximates spatially varied parameters with a constant value. Although the form is simpler, it loses detailed information of spatial structure. This is the reason why we need the agent-based simulation above rather than numerical analysis based on approximated equations in a network.

Under the approximation that the probability that the state of node 0 is constant $p0$ (mean field approximation and steady state), the following steady state probability of $p0$ is obtained. For the calculation of the steady state probability, refer to Sect. 7.3.

$$p0 = \frac{P_{ra}\ (2 - 2P_r + P_r\ P_{ra})}{P_r\ (1 - P_{ra})^2}$$

Fig. 3.5 The number of normal nodes after 800 steps plotted when the successful repair probability P_{ra} varies. The three lines correspond to P_r: 0.3, 0.6, 1.0 from *left* to *right*

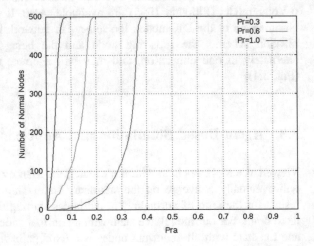

This steady state probability also matches qualitatively with the above simulation results.

When P_{ra} increases, the number of normal nodes rapidly increases. The steeper the curve, the smaller the probability P_r. That is, when P_{ra} is less than 0.4, P_r should be small and hence repair should be done infrequently.

Further, there is a critical point of P_{ra} that must be exceeded to clean up the network. The critical point becomes larger as the probability P_r becomes larger. The critical points are plotted in Fig. 3.6. The left region is the *frozen* region where all the nodes become normal, and the right region is the *active* region where some abnormal nodes remain in the network.

When we use both mean field approximation and agent based computer simulation, we can draw two borderlines. From *p0* above, we get

$$P_{ra} = 1/2P_r,$$

which means that the condition for eradicating abnormal nodes under the mean field assumption is: the repair success rate by abnormal nodes must exceed half of the repair rate.

This borderline qualitatively matches the borderline obtained by the computer simulation but quantitatively shifted downward with P_r almost 0.2 where P_r is not close to 0 (Fig. 3.7). This means that the above condition is just a sufficient (not necessary) condition only when P_r is not close to 0. From the approximated equation of *p0* above, we know that the point when $P_r = 0$ is a singular point. From a modeling point of view, it is also singular because when $P_r = 0$, the abnormal nodes remain abnormal, and hence the ratio of abnormal nodes remains constant all the time after the initial setting of normal nodes and abnormal nodes, for there is no repair at all.

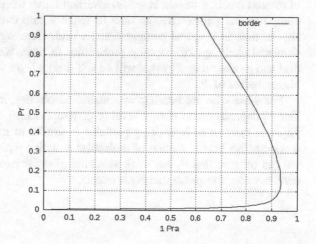

Fig. 3.6 Frozen phase (*left region* where all the nodes are normal) and active phase (*right region* where some nodes remain abnormal)

Fig. 3.7 The theoretical
(MFA) and computer
simulated borderlines
separating the frozen phase
(*left region* where all the
nodes are normal) and active
phase (*right region* where
some nodes remain abnormal)

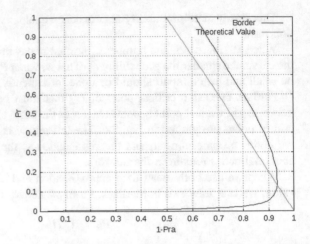

The computational and modeling challenge exists where P_r is small (close to zero). Generally, scaling issues are fundamental even in physical (reduced scale) modeling: when one builds a reduced size physical model for the analysis of, for example, fluid dynamics, it is not possible to make everything small such as particles of air or water. If we call this problem the physical and spatial scaling problem, then the problem we are facing here may be called a probabilistic and temporal scaling problem.

As the parameter P_r becomes smaller, computer simulation becomes harder because it takes much longer. The Monte Carlo simulation, for example, depends on virtually created (with computations) probabilistic events. Agent-based simulations (also called multi-agent simulations) may also depend on probabilistic events. From a modeling perspective, it can be formalized as how to make the point of interest (such as threshold value) invariant under the re-scaling of P_r, and from a computational perspective, as how to make a rare event efficiently happen in a shorter time in a virtual computational world. Here, again we encounter the fundamental challenge of mapping probability to time. Nevertheless, in Chap. 7 we find and propose a duality that will be self-dual when P_r is small enough to neglect higher orders of P_r.

For those who are familiar with statistical physics, it may be easier to imagine that P_r depends on an energy level (e.g., temperature). Raising the temperature in physical systems of interacting particles is similar to raising P_r in self-repair networks: active kinetic motion of molecules to active repairing of nodes. In Fig. 3.8, we can observe that P_r can work as a kind of order parameter: when P_r exceeds a certain value, the phase transition occurs.

Fig. 3.8 Threshold values of a phase transition can be observed: when P_r is raised while keeping P_{ra} fixed, abnormal nodes appear after the threshold value is exceeded

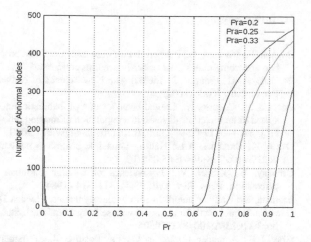

3.6 Conclusion

As a first step toward an adaptive defense of the network, we studied a simple problem of cleaning up the network by mutual copying. In a probabilistic framework, a model similar to the Domany-Kinzel model can be obtained. A critical phenomenon is observed: when copying from faulty nodes has a certain rate of infecting rather than cleaning, then mutual copying could spread contamination; hence the rate should be carefully identified and controlled.

We presented a naive model for a naive problem of cleaning networks by mutual copying. When applied to existing information networks such as the Internet and sensor networks, self-repair networks should deal with not only static regular networks but also dynamic growing networks such as Barabási's scale-free networks, and should deal with not only synchronous repair but also asynchronous repair.

Although threshold values of parameters for cleaning networks by mutual copying have been obtained by computer simulations, they should be determined with mathematical rigor. Also, theoretically the model and results here should be related to the cellular automaton of Gacs, but this problem is not addressed in this book. Analogy to the immune system would suggest that nodes not only repair by copying (effector counterpart of the immune cells) but also recognize abnormal nodes (receptor counterpart).

We have shown that even a uniform and non-strategic repair could clean all nodes of the self-repair network within a certain parameter region (frozen region). However, it is obviously inefficient and a waste of resources to repair any nodes at any time, for the environment (e.g., fault rate) can vary from node to node and time to time. In the following Chap. 4, we will introduce strategic repair where cooperation (repair) and defection (not repair) can be switched depending on the strategy of neighbor nodes.

References

Bagnoli, F., Boccara, N., Palmerini, P.: Phase transitions in a probabilistic cellular automaton with two absorbing states (1997). arXiv preprint cond-mat/9705171

Barabási, A.-L., Frangos, J.: Linked: The New Science of Networks Science of Networks. Basic Books, New York (2002)

Boccara, N., Cheong, K.: Critical-behavior of a probabilistic-automata network SIS model for the spread of an infectious-disease in a population of moving individuals. J. Phys. A Math. Gen. **26** (15), 3707–3717 (1993). doi:10.1088/0305-4470/26/15/020

Dezso, Z., Barabási, A.L.: Halting viruses in scale-free networks. Phys. Rev. E **65**(5) (2002). doi:10.1103/PhysRevE.65.055103

Domany, E., Kinzel, W.: Equivalence of cellular automata to Ising models and directed percolation. Phys. Rev. Lett. **53**(4), 311–314 (1984)

Eloranta, K.: Cellular automata for contour dynamics. Physica D **89**(1), 184–203 (1995)

Gacs, P.: Reliable cellular automata with self-organization. J. Stat. Phys. **103**(1–2), 45–267 (2001). doi:10.1023/A:1004823720305

Gray, L.F.: A reader's guide to Gacs's "Positive Rates" paper. J. Stat. Phys. **103**(1–2), 1–44 (2001). doi:10.1023/A:1004824203467

Ishida, Y.: Immunity-Based Systems: A Design Perspective. Springer, New York Incorporated (2004)

Ishida, Y.: A critical phenomenon in a self-repair network by mutual copying. In: Knowledge-Based Intelligent Information and Engineering Systems, pp. 86–92. Springer, Berlin (2005)

Kinzel, W.: Phase transitions of cellular automata. Zeitschrift für Physik B Condensed Matter **58** (3), 229–244 (1985)

Rhodes, C.J., Anderson, R.M.: Dynamics in a lattice epidemic model. Phys. Lett. A **210**(3), 183–188 (1996). doi:10.1016/S0375-9601(96)80007-7

Rhodes, C.J., Anderson, R.M.: Forest-fire as a model for the dynamics of disease epidemics. J. Franklin Inst. **335B**(2), 199–211 (1998). doi:10.1016/S0016-0032(96)00096-8

Vichniac, G., Tamayo, P., Hartman, H.: Annealed and quenched inhomogeneous cellular automata (INCA). J. Stat. Phys. **45**(5–6), 875–883 (1986)

Chapter 4
Controlling Repairing Strategy: A Spatial Game Approach

Abstract We address the problem of cleaning up a contaminated network by mutual copying. This problem involves not only the double-edged sword where copying could further spread contamination but also mutual cooperation where resource-consuming copying could be left to others. This chapter applies the framework of the "spatial prisoner's dilemma" in an evolutionary mechanism, with the aim of appropriate copying strategies emerging in an adaptive manner to the network environment. To introduce costs for repairing, an agent-based simulation is used to express cost as the resources consumed by each agent. As the benefit of repair to compensate for the cost, repairing is associated with copying the strategy of the repairing agents to the repaired agents. This biological mechanism for maintaining cooperative (and repairing) agents is the subject of this chapter. The risk of repairing by copying (another edge of the sword) is implemented as an increase of the failure rate of the repaired agent.

Keywords Strategic repair · Spatial prisoner's dilemma · Spatial strategies · Maintenance of cooperating clusters · Evolutionary mechanisms · Agent-based simulations

4.1 Introduction

Now we are ready to design a framework for selfish machine communities (whose actions are, however, restricted to repair for modeling purposes). Naturally, we must consult game theory which also has a theory of mechanism design based on selfish agents. Although we discussed incentive (Chap. 2) and macroscopic features

Most results of this chapter are presented in Ishida and Mori (2005a).

© Springer International Publishing Switzerland 2015
Y. Ishida, *Self-Repair Networks*, Intelligent Systems Reference Library 101,
DOI 10.1007/978-3-319-26447-9_4

(Chap. 3) of involving repair at the node (agent) level, we need to design the framework considering how, when, where the repair should be done assuming selfish agents. We therefore use the popular "prisoner's dilemma" game by assigning cooperation and defection to repair and not-repair.

Frank (1998) noted three "exchange rates" that can be influenced by natural selection in his economic theoretical foundation of social evolution. One is "the coefficient of relatedness from kin selection theory," in addition to other standard scaling factors of time value scaling to compare the present offspring value and the future offspring value, and marginal value scaling to compare costs and benefits. In this regard, relatedness in the self-repair network is expressed by closeness in the network, which is visualized by an edge in a graph or neighbors in a lattice. Close nodes in a network share common interests and risks. This chapter uses a square lattice for visualizing connected components (clusters).

Many artificial networks such as the Internet have been shown to be a *scale-free* network (Barabási and Frangos 2002). This finding has had an impact on many researches such as on the security of the network (Dezso and Barabási 2002). On the other hand, self-repair systems including those inspired by the immune system have been studied extensively (Ishida 2004). The adaptive nature of the immune system to the environment is of interest, which is realized by cooperative working among inhomogeneous agents with distinct strategies. Further, a new idea of recovery-oriented computing has been proposed (Brown and Patterson 2001). We have studied the problem of cleaning up the network by mutual copying, and proposed a self-repair model (Chap. 3) that can be equated with probabilistic cellular automata (Domany and Kinzel 1984). The model considers only synchronous and non-selective copying (overwriting, or modifying the state in general). This chapter, however, focuses on adaptive copying using the Prisoner's Dilemma approach (Axelrod 1987, 1984; Axelrod and Dion 1988; Axelrod and Hamilton 1981).

In the case of self-repair systems with autonomous distribution, abnormal agents may adversely affect the system when they try to repair other normal agents. Additionally, because repairing uses some resources, frequent repairs reduce the performance of the system. Although the frequency of repairs has to be decided in consideration of the system environment, it cannot be easily determined because the environment within a system changes with time.

This chapter proposes a model where the agents (nodes) select a strategy depending on the system environment. In this model, an agent is assigned some resources and a strategy, and it repairs according to this strategy. We aim to raise the reliability of the system without reducing the performance, in a fully distributed framework.

When the self-repair is done in an autonomous distributed manner, each agent does not voluntarily repair other agents to save its own resources, thus leaving many faulty agents not repaired. This situation is similar to the dilemma that occurs in the Prisoner's Dilemma. Thus, we use an approach of a distributed version (Matsuo 1989) or spatial version (Grim 1995; Nakamaru et al. 1998; Nowak and May 1992) of the Prisoner's Dilemma for the emergence of cooperative collectives and for controlling copying to save resources.

The problem tackled in this chapter is how the self-repairing is regulated with the spatial strategy when repair costs are involved:

If the self-repairing involves cost and a controlling strategy, how can the self-repairing be controlled?

In the modeling of this chapter, not only a game theoretic framework of strategy but also an evolutionary framework of "survival of the fittest" and mutation is incorporated. The remaining resource is used to measure the payoff to determine the fittest. To observe the effect of control, several constant parameters related to resource consumption are included in the model assumptions:

- Fault: Each node becomes abnormal independently and randomly with a constant failure rate, but the failure rate increases with a constant damage rate when the node is repaired by abnormal nodes. Although normal nodes have a constant amount of resource, abnormal nodes do not.
- Repair: Repair by normal nodes targets only abnormal nodes, makes them normal, and refreshes the failure rate to the base value; however, repair by abnormal nodes targets both normal and abnormal nodes and raises their failure rates when the repair succeeds. Repair by normal nodes consumes a constant amount of resource of the repairing node, whereas repair by abnormal nodes is done without any consumption of resources.
- Control: Repair actions are determined based on the spatial pattern of the actions in the neighbor (spatial strategy). Once a repair action is determined, it is executed asynchronously and probabilistically with a constant rate (repair rate). Repair can be done on multiple nodes simultaneously in the neighbors as long as their resource consumption does not exceed the constant limit (available maximum resource).
- Strategy: The strategy is updated in a constant period (strategy update cycle). In updating, the strategy of the neighbor (eight-neighbor) nodes that earned the highest payoff is copied. Payoff is measured by the available resource owned by the node when updating the strategy. In the strategy updating, strategy may mutate probabilistically with a constant rate (mutation rate). In repairing, the strategy of the repairing node is copied to the repaired node (strategy copying in repair).
- Network: An $L \times L$ two-dimensional square lattice with eight neighbors and with the periodic boundary condition.

The assumption of strategy copying in repair plays a crucial role in an evolutionary mechanism design, for it amounts to not only fixing the functional level but also to spreading the strategy by directly increasing the "fitness" of the strategy in an evolutionary context.

4.2 Spatial Strategies in SPD

The spatial version of the Prisoner's Dilemma (PD) has been extensively studied (e.g., Nowak and May 1992; Grim 1995). We considered a spatiotemporally generalized strategy, and proposed a generalized TFT (Tit-for-Tat) such as $k1C$, $k2D$ and their combination $k1C$-$k2D$, where $k1$ is a parameter indicating generosity and $k2$ contrariness (Ishida and Mori 2005b). The dynamics of these spatial strategies in a two-dimensional lattice have been also studied in a noisy environment.

The PD is a game played just once by two players with two actions (cooperation, C, or defection, D). Each player receives a payoff (R, T, S, P) where $T > R > P > S$ and $2R > T + S$.

In IPD, each player (and hence the strategy) is evaluated. In the SPD, each site in a two-dimensional lattice corresponds to a player. Each player plays the PD with the neighbors, and changes its action according to the total score it received.

Our model generalized the SPD by introducing spatial strategy. Each player is placed at each cell of the two-dimensional lattice. Each player has an action and a strategy, and receives a score. The spatial strategy determines the next action dependent upon the spatial pattern of actions in the neighbors. Each player plays the PD with the neighbors, and changes its strategy to the strategy that earns the highest total score among the neighbors. Table 4.1 is the payoff matrix of the PD. In our simulations, R, S, T, and P are respectively set to 1, 0, b $(1 < b < 2)$ and 0 in the simulations shown below following Nowak-May's simulations (Nowak and May 1992).

Table 4.1 The payoff matrix of the Prisoner's Dilemma game. R, S, T, P are payoffs to player 1 $(1 < b < 2)$

Player 1	Player 2	
	C	D
C	R (1)	S (0)
D	T (b)	P (0)

Our SPD is done with spatial strategies: the next action will be determined based on the pattern of neighbors' actions. The score is calculated by summing up all the scores received from the PD with eight neighboring players. After r (strategy update cycle) steps of interactions with neighbors, the strategy with the highest score among the neighbors will be chosen. Thus, the strategy will be updated every r steps. In an evolutionary framework, the strategy will be also changed by a mutation rate where mutation operates on the string of the strategy code as follows.

To specify a spatial strategy, actions of the eight neighbors (the *Moore* neighbors) and the player itself must be specified (Fig. 4.1), hence 2^9 rules are required. For simplicity, we restrict ourselves to a "totalistic spatial strategy" that depends on the number of D (defect) actions of the neighbor, not on their positions.

To represent a strategy, a bit sequence is used (Matsuo 1989) whose lth element is C (D) if the action C (D) is taken when the number of D of the neighboring players is l ($l = 0, 1, \ldots, 8$). For example, *All-C* is [CCCCCCCCC] and *All-D* is [DDDDDDDDD]. The number of strategies is $2^9 = 512$. As a typical strategy, we define kD that takes D if $l > k$ and C otherwise. For example, *2D* is [CCDDDDDDD]. This kD strategy can be regarded as a spatial version of TFT where k seems to indicate the generosity (called *spatial generosity*) (how many D actions in the neighbors are tolerated).

Strategy kC can be similarly defined: it takes C if $l > k$ and D otherwise. For example, *4C* is [DDDDCCCCC] as shown in Fig. 4.1. The number k in the strategy kC seems to indicate the contrariness that some players would cooperate even if k players in the neighbor defect. Further, $k1D$-$k2C$ is a combination of $k1D$ and $k2C$. For example, *2D-7C* is [CCDDDDDCC]. In our previous studies, the action error has an effect of favoring more generous spatial strategies in the sense that more D actions should be forgiven in the neighbor's actions (Ishida and Mori 2005b). The effects of errors have been already studied even by repeated PD (Boyd 1989).

Fig. 4.1 A strategy code for spatial strategies

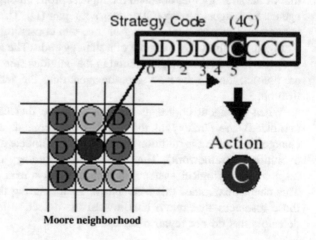

Strategy Code   (4C)

DDDDCCCCC
0 1 2 3 4 5

Action
C

Moore neighborhood

4.3 A Model Incorporating SPD

The agents in the system are assigned tasks that they should do. The agents are assumed to use corresponding resources while repairing, making it difficult for them to perform much of the tasks. The agents have to do the tasks assigned to them, but without doing repair, abnormal agents increase and hence the performance at the system level decreases, causing a dilemma. Our model adopts the spatial version of the Prisoner's Dilemma (SPD) approach. The SPD model arranges players that interact with neighboring players in a two-dimensional lattice space. Each agent is assumed to use *spatial strategies* (Ishida and Mori 2005b).

Modeling for the agent-based simulation allows us to introduce realistic parameters for engineering, although the realistic approach prevents us from conducting a mathematical investigation at the same time. The first difference from the model in Chap. 3 is that the model here involves a cost for repairing. The asymmetric nature that the cost is imposed on the repairing agent while the benefit of the repairing accrues to the entire system naturally biases each agent's strategy toward defection, i.e., not repairing.

The agent becomes abnormal depending on the failure rate λ. Each agent has a strategy code and repairs according to that code. Repairing is done depending on the repairing rate γ and a repaired agent becomes normal. Also the agent uses the quantity of repair resource $R\gamma$ for repairing. The agent is able to repair more than one agent, provided that the quantity of maximum resource $Rmax$ is not exceeded. We call the resource that is not used for repairing the available resource Ra, and consider it as the *score* of an agent. If an abnormal agent repairs another agent, the failure rate λ of the repaired agent is increased by damage rate δ. This damage by abnormal nodes represents another edge of the double-edged sword and plays a similar role to the repair success rate by abnormal nodes Pra in the model of Chap. 3. Through the strategy update cycle r, the strategy of all agents is updated to that of the strategy that obtained the highest profit among the *Moore* neighborhood agents. Each agent repairs (C) or does not repair (D). The strategy code of the agent is expressed as a character string of nine bits consisting of Cs and Ds.

Figure 4.1 shows an example of a strategy code. The agent counts the number of D agents in its *Moore* neighborhood in the previous step. In this case, five D agents are present; then in the next step, the agent does the action (C) represented by the fifth bit.

When the agent copies its content (software that can be copied and be contaminated), the strategy of the agent is also copied. Thus, the strategy will be changed at copying in addition to every strategy update cycle (and at mutation in an evolutionary framework). This association of strategy copying when repairing is inspired by biological systems in which genes spread associated with offspring. This chapter examines this mechanism, for measuring the score of the strategy by those resources that retain bias toward the defection-favored direction (favoring defectors that do not repair other agents).

This chapter restricts strategies to the kC strategy, which is composed of only nine among the 512 existing strategies. In the kC strategy, the agent does the action of D if the number of D agents is less than k, and it does the action of C if the number of D agents is greater than or equal to k. The kC strategies are rather unnatural, but we restrict ourselves to them to compensate for the defector-favoring bias noted above.

4.4   Agent-Based Simulations with a Square Lattice

The simulation is done in a two-dimensional lattice space with an agent existing in each cell. Throughout the simulations described in this section, the parameters are: failure rate 0.001, repair rate 0.01, quantity of maximum resource 8, quantity of repair resource 2, damage rate 0.1 and strategy update cycle 200.

Figure 4.2 plots the number of failure agents (average number of 10 independent trials) with damage rate varying from 0.00 to 0.30. In this simulation, all the agents are set to be normal initially. It can be observed that there is a threshold between 0.12 and 0.13 above which the number of failure agents explodes. The threshold is also observed in a somewhat different but similar (and simpler) model for self-repair (Chap. 3).

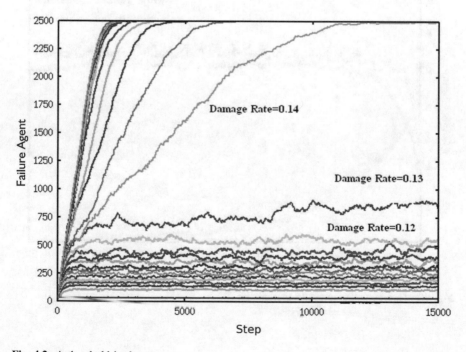

Fig. 4.2 A threshold in damage rate

To examine the efficacy of the strategic repair, we compare it with two trivial strategies: the one where the agents always repair, and the one where the agents never repair. We use the same parameters for these three strategies with the damage rate fixed at 0.1. To observe the effect of repair, 50 % of the agents comprising the 50 × 50 space are randomly selected and set to be abnormal in the initial configuration.

Figure 4.3 shows the available resource (averaged over 40 independent trials) in each model. We can observe that both *the always repair* and *the strategic repair* become stable after some number of steps (an oscillation observed in the strategic repair is due to the change of numbers in strategies at the strategy update cycle). However, more available resource remains in *the strategic repair*. Moreover, *the strategic repair* becomes stable twice as fast.

Figure 4.4 shows the number of agents in each strategy of *the strategic repair*. We can see that when the number of abnormal agents increases, agents with easily repairing strategies like *0C* also increase. Otherwise, agents using hardly repairing strategies like *8C* increase. Therefore, the agents choose an appropriate strategy depending on the environment.

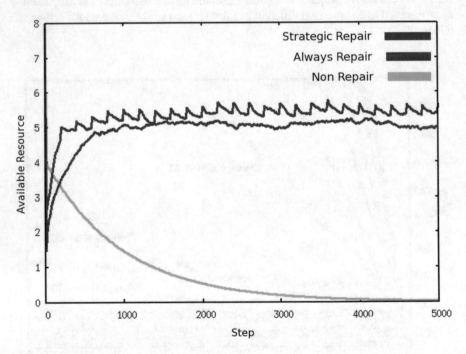

Fig. 4.3 Time evolution of available resources

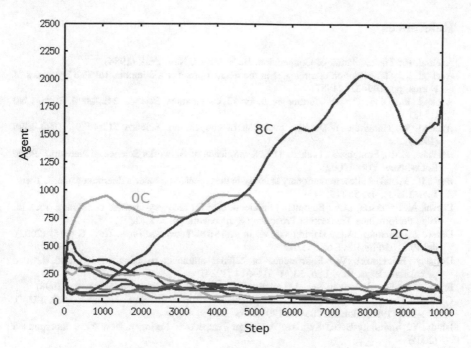

Fig. 4.4 Time evolution of the fraction of agents with spatial strategies

4.5 Conclusion

We have studied the problem of when can a network be cleaned up by mutual copying. We proposed a probabilistic cellular automaton model for synchronous copying for simplicity (Chap. 3). In this chapter, a game theoretical approach of the spatial Prisoner's Dilemma was used. With this approach, mutual copying can be controlled so that agents will mutually repair by copying but not too excessively such that their resources are exhausted. It is also shown that a strategy appropriate for the environment will emerge when the system environment changes.

However, to compensate for the asymmetric nature of the cost and benefit of repairing in self-repair networks, we need to introduce a mechanism of a biological (gene-like) spreading of the repairing strategy, as well as to restrict the strategy space to kC.

In the following Chap. 5, we will observe that strategic repair is able to adapt to a spatially heterogeneous environment (e.g., the fault rate varies from node to node) and to a temporally dynamic environment (e.g., the fault rate changes from time to time).

References

Axelrod, R.: The Evolution of Cooperation. Basic Books, New York (1984)

Axelrod, R.: The evolution of strategies in the iterated prisoner's dilemma. In: The Dynamics of Norms, pp. 199–220 (1987)

Axelrod, R., Dion, D.: The further evolution of cooperation. Science **242**(4884), 1385–1390 (1988)

Axelrod, R., Hamilton, W.D.: The evolution of cooperation. Science **211**(4489), 1390–1396 (1981)

Barabási, A.-L., Frangos, J.: Linked: The New Science of Networks Science of Networks. Basic Books, New York (2002)

Boyd, R.: Mistakes allow evolutionary stability in the repeated prisoner's dilemma game. J. Theor. Biol. **136**(1), 47–56 (1989)

Brown, A., Patterson, D.A.: Embracing failure: A case for recovery-oriented computing (roc). In: High Performance Transaction Processing Symposium, pp. 3–8 (2001)

Dezso, Z., Barabási, A.L.: Halting viruses in scale-free networks. Phys. Rev. E **65**(5) (2002). doi:10.1103/PhysRevE.65.055103

Domany, E., Kinzel, W.: Equivalence of cellular automata to Ising models and directed percolation. Phys. Rev. Lett. **53**(4), 311–314 (1984)

Frank, S.A.: Foundations of Social Evolution. Princeton University Press, Princeton (1998)

Grim, P.: The greater generosity of the spatialized prisoners-dilemma. J. Theor. Biol. **173**(4), 353–359 (1995). doi:10.1006/Jtbi.1995.0068

Ishida, Y.: Immunity-Based Systems: A Design Perspective. Springer, New York Incorporated (2004)

Ishida, Y., Mori, T.: A network self-repair by spatial strategies in spatial prisoner's dilemma. In: Knowledge-Based Intelligent Information and Engineering Systems, pp. 79–85. Springer, Berlin (2005a)

Ishida, Y., Mori, T.: Spatial strategies in a generalized spatial prisoner's dilemma. Artif. Life Robot. **9**(3), 139–143 (2005b)

Matsuo K., Adachi, N.: Metastable antagonistic equilibrium and stable cooperative equilibrium in distributed prisoner's dilemma game. Paper presented at the international symposium systems research, informatics and cybernetics, Barden-barden (1989)

Nakamaru, M., Nogami, H., Iwasa, Y.: Score-dependent fertility model for the evolution of cooperation in a lattice. J. Theor. Biol. **194**(1), 101 (1998)

Nowak, M.A., May, R.M.: Evolutionary games and spatial chaos. Nature **359**(6398), 826–829 (1992)

Chapter 5
Adaptive Capability in Space and Time

Abstract In an information network composed of selfish agents pursuing their own profits, undesirable phenomena such as spam mail occur if the profit sharing and other game structures permit such equilibriums. We focused on applying the spatial Prisoner's Dilemma to control a network of selfish agents by allowing each agent to cooperate or to defect. Cooperation and defection respectively correspond to repair (using self-resource) and not repair (thus saving resource) in a self-repair network. Without modifying the payoff of the Prisoner's Dilemma, the network will be absorbed into the state of Nash equilibrium where all the agents become defectors and abnormal. Similarly to *kin selection*, agents favor survival of neighbors in organizing these two actions to prevent the network from being absorbed if payoffs are measured by summing all the neighboring agents. In this chapter, using the agent-based simulation, we assert that, even with this modification, the action organization exhibits spatial and temporal adaptability to the environment.

Keywords Adaptation to the environment · Dynamic environment · Spatial strategies · Maintenance of cooperating clusters · Control of repair rate · Agent-based simulations

5.1 Introduction

An old Japanese saying goes, "Mercy is not for the people" (literal translation), which means that helping other persons is helping yourself. To prevent nodes from being not repairing (not helping) and hence from being the macro state of all abnormal, whereas Chap. 4 tested a biological mechanism of spreading the repairing strategy (genes) directly associated with repairing (reproduction), this chapter tests the social mechanism of extending the reward-receiving unit.

Most results of this chapter are presented in Ishida and Tokumitsu (2008), Oohashi and Ishida (2007).

© Springer International Publishing Switzerland 2015   59
Y. Ishida, *Self-Repair Networks*, Intelligent Systems Reference Library 101,
DOI 10.1007/978-3-319-26447-9_5

We have introduced an (evolutionary) game theoretic framework, and discussed the possibility of controlling repair actions by using the spatial prisoner's dilemma and by extending the strategy from the conventional temporal (time-based) strategy to the spatial strategy (spatial configuration based). We would naturally expect the adaptive feature of the evolutionary game theoretic design. We believe that spatial (or configuration based) extension of stability (ESS) and equilibrium (Nash equilibrium) are needed for a rigorous analysis of spatial strategy, which will be left for Chap. 6. Here, we restrict ourselves to just two strategies: All-C and All-D for simplicity of studying adaptability.

The Prisoner's Dilemma (PD) has motivated studies on why cooperation emerges in many autonomous systems with selfish agents. The spatial Prisoner's Dilemma (SPD) (Nowak and May 1992; Nowak 2006) (Sect. 4.2), in turn, has led to studies on why clusters of cooperators emerge when selfish agents interact with a spatially restricted world. We noted that the spatial Prisoner's Dilemma may be used for organizing an autonomous and distributed network if cooperation and defection are properly mapped to mutual operations among agents.

Spam mail is a phenomenon that can be understood as agents implementing a defective strategy. As shown by such a case, the SPD or the game theoretic approach in general requires some modification to be applied to information networks such as the Internet. A game theoretic approach assumes that each player will sense, decide, act, and receive the reward/penalty as a distinct identity. The first modification would be to relax these assumptions. This chapter focuses on relaxing the last one: each player receives the reward (positive or negative) not exclusively but can share it with neighbors. Indeed, virus infection and intrusion to neighbor nodes (where an agent is placed at each node) is a crucial concern for most nodes. Sharing the reward/penalty with neighbor nodes amounts to generalizing the unit of reward-receiving. This wider scope of reward-receiving would have a similar effect on the evaluation of profit over a longer time span, and hence space-time interplay is an interesting issue; however, this interplay will be discussed in Chap. 6.

As for mapping cooperation/defection to control operations, this book focuses on only repair/not-repair in a self-repair network (Oohashi and Ishida 2007; Tokumitsu and Ishida 2008). However, there are many ways of mapping to deal with other issues of autonomous distributed networks, such as mapping to sharing resources/not sharing; and communicating messages/not communicating. It is also interesting to map the free-rider in the public goods game to not-repair rather than the defector in the prisoner's dilemma.

This chapter focuses on the adaptive capability of spatial control of self-repair:

Can self-repairing controlled by spatial strategy adapt to the network environment?

In the modeling of this chapter, to express the network environment, failure rate is changed spatially and temporally. The most important difference from the previous chapter is that the mechanism is a *social* one not a *biological* one, that is, the payoff is extended to include not only the remaining resource of the node but also those of the neighbor nodes (extended payoff, called *systemic payoff*). This extended payoff is used instead of the strategy copying in repair of the *biological* mechanism.

- Fault: Each node becomes abnormal independently and randomly with a constant failure rate, but the failure rate increases with a constant damage rate when the node is repaired by abnormal nodes. Although normal nodes have a constant amount of resource, abnormal nodes do not.
- Repair: Repair by normal nodes targets only abnormal nodes, makes them normal, and refreshes the failure rate to the base value; however, repair by abnormal nodes targets both normal and abnormal nodes and raises their failure rates when the repair succeeds. Repair by normal nodes consumes a constant amount of resource of the repairing node, whereas repair by abnormal nodes is done without any consumption of resources.
- Control: Repair actions are determined based on the spatial pattern of the actions in the neighbors (spatial strategy), but here only All-C and All-D are involved as spatial strategies. Once a repair action is determined, it is executed asynchronously and probabilistically with a constant rate (repair rate). Repair can be done on multiple nodes simultaneously in the neighbors as long as their resource consumption does not exceed the constant limit (available maximum resource). Repair actions are flipped with a constant rate (action error rate).
- Strategy: The strategy is updated in a constant period (strategy update cycle). In updating, the strategy of the neighbor (eight-neighbor) nodes which earned the highest payoff is copied. Payoff is measured by the available resource owned by the neighbor nodes as well as the node itself (systemic payoff) when updating the strategy. In the strategy updating, strategy may mutate probabilistically with a constant rate (mutation rate). Because the strategies are limited to All-C and All-D, incorporating the mutation rate is virtually equivalent to incorporating the action error rate here.
- Network: An $L \times L$ two-dimensional square lattice with eight neighbors and with the periodic boundary condition.

The mechanism of extended payoff is incorporated to make the cooperation (repairing the neighbor nodes) survive. The mechanism of strategy copying in repair is replaced with the mechanism of extended payoff, for the former might have an evolutionary impact other than fixing normal functions. However, which is better for realizing the self-repair network in practical systems is another question.

The differences between the tested mechanisms of Chap. 4 and this chapter can be summarized as:

- Chapter 4: Spatial strategies involving copying the strategy of the repairing agent when repairing;
- In this chapter: Extended payoff incorporating not only its own remaining resources but also all the resources in the neighbor.

Whereas the spatial strategies kC were tested in Chap. 4, this chapter is limited to only two strategies of All-C and All-D in strategic repair for simplicity in analyzing the adaptive nature of strategic repair. Actually, All-C (0C) and All-D (0D) are not spatial strategies, for they do not take into consideration the spatial configuration of neighbors. Therefore, the spatial strategy kD will be investigated in Chap. 6.

5.2 Basic Model and Assumptions

We consider the problem of cleaning up a network composed of autonomous agents with binary state: normal or abnormal. Network cleaning is carried out by mutually repairing agents. Repairing is done by each agent transferring its own content to other agents. We assume that the state of an agent (normal/abnormal) cannot be known by the agent itself or by the other agents. Thus, repairing by abnormal agents could harm other agents rather than repairing them, thus mutual repairing involves the double-edged sword aspect: an inappropriate strategy would lead to further contamination of the network. Furthermore, repairing uses resources of the repairing agent, and so whether to repair should be decided according to the resources used and resources remaining in the system and the network environment.

Selfish agents will determine their actions by considering their payoffs. To implement this selfish framework, we investigate a game theoretic approach. Since repair/not-repair corresponds to cooperate/defect in the SPD preserving the structure of the payoff matrix, we will use the SPD to organize the actions of agents. Throughout this book, we will restrict ourselves to two actions, repair and not repair, in the self-repair network. Repairing actions are carried out asynchronously by neighboring agents. Every agent has its own strategy that will determine its actions.

The SPD has been studied to investigate when, how, and why cooperation emerges among selfish agents when they are spatially arranged, hence interactions are limited only to their neighbors (Nowak and May 1992). Each player is placed at each cell of the two-dimensional lattice. Each player has an action and a strategy, and receives a score. Each player plays the Prisoner's Dilemma (PD) with the neighbors, and changes its strategy to the strategy that earns the highest total score among the neighbors. We will use this deterministic SPD.

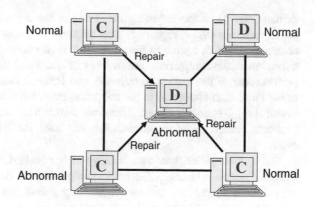

Fig. 5.1 A neighborhood of the self-repairing network whose actions are organized by SPD (*All-C and All-D only*)

In sum, our model has the following assumptions (Fig. 5.1):

- Each agent has a binary state: normal or abnormal; and a binary strategy: always cooperate (All-C) or always defect (All-D).
- Agents with the All-C strategy (cooperators or C) will repair the agents in the neighbor, whereas agents with the All-D strategy (defectors or D) will not repair.
- Each agent becomes abnormal with a certain failure rate.
- Each agent is given a fixed amount of resources updated in a unit of time.
- Repairing requires some fixed amount of resources of the repairing agents.
- Repairing (and hence state updating) is done asynchronously.

In the stochastic version, the agent will decide its action based on a probability proportional to the difference between its own payoff and the highest payoff in the neighbors' agents [similarly to replicator dynamics (Taylor and Jonker 1978; Hofbauer and Sigmund 2003)].

5.3   Agent-Based Simulations in Heterogeneous and Dynamic Environments

Every agent is placed at each cell of a two-dimensional lattice with a periodic boundary condition (and so the lattice size coincides with the number of agents), and every agent will interact with the agents in the neighboring cells. In this study, the *Moore* neighborhood (eight neighboring agents) is used.

Simulations are conducted using the following parameters. Each agent has a failure rate (λ). This failure rate will be varied to implement spatially and temporally varied environments. The repair will be done by a repair rate (α), and the successfully repaired agents are made normal. The adverse impact in repairs caused by abnormal agents is implemented by raising the failure rate (by the amount of damage rate δ) of the target agents. Further, the agents are assumed to consume some resources (R_λ) in repairing. This amounts to a cost for cooperation, and hence encourages selfish agents to free-ride. The agents have to do the tasks assigned to them; but without repairing, the number of abnormal agents increases causing the performance of the system to degrade, and hence a dilemma. The agent is able to repair more than one agent in the neighbor, provided that the quantity of maximum resource R_{max} is not exceeded. Abnormal agents have no resource. We consider the available resource (the resource that is not used for repairing) as the score of an agent.

The agents update their own strategy after r (called *strategy update cycle*) steps from the previous change. The next strategy will be chosen from the strategy that earned the highest score among neighboring agents. The total available resource is updated in each step by adding a fixed resource and subtracting the consumed resource from the current resource value. Table 5.1 lists the simulation parameters used in this study.

5.3.1   *Repair Rate Control with Systemic Payoff*

Here we compare the case when the payoff is evaluated as usual and that when the payoff is evaluated collectively in the neighborhood. Without the collective evaluation, defectors (All-D) will eradicate cooperators (All-C) (Fig. 5.2b); hence all the agents will remain silent without repairing any agents. Thus, eventually all the agents will become abnormal with a positive failure rate (Fig. 5.2a).

We devised a payoff to prevent all the agents from taking D actions and from being abnormal by incorporating not only their own remaining resources but also all the resources in the neighborhood. This modified payoff (called *systemic payoff*) has an impact on making agents more attentive by caring for neighboring agents that might be able to repair the agents in the future. This modification reminds us of *kin selection* theory (Hamilton 1964; Smith 1964) that emphasizes the survival of close

Table 5.1 List of parameters for simulations

	Description	Value
$L \times L$	Size of the space	50×50
N	Number of agents	2500
$N_f(0)$	Initial number of abnormal agents	100
λ	Failure rate	0.01
α	Repair rate	0.1
δ	Damage rate	0.1
r	Strategy update cycle	100
R_{max}	Maximum resource	25
R_λ	Resource used for repairing	1

Fig. 5.2 SPD with simple payoff measured by available resources of the agent (Oohashi and Ishida 2007). Parameters are: failure rate 0.005–0.10, repair success rate 0.1, damage rate 0.1, strategy update cycle 20, maximum resources 9, cost for repair 1. 100 randomly-chosen agents are made abnormal and a randomly-chosen half of agents take All-D initially

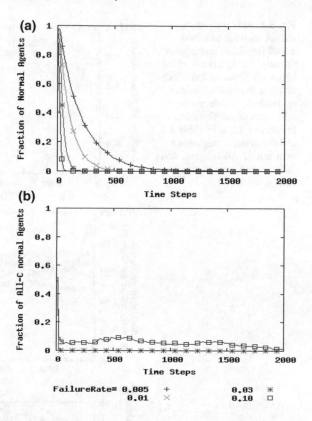

relatives. As already noted, the neighbor nodes are important in information networks, since security issues (infection and intrusion) and the performance of the neighbor nodes are critical to every node.

Since abnormal nodes have no resources at all and the score of each agent is the total of the remaining resources in the neighbor nodes including the agent itself, the systemic payoff rewards the agent surrounded by normal agents and punishes the agent surrounded by abnormal agents.

Simulations are conducted for strategic repair with modified payoff. Figure 5.3 plots the time evolution of the fraction of normal agents (a), available resources left in the system (b), and the fraction of agents with All-C (c).

It can be seen that strategic repair with modified payoff is capable of selecting an appropriate strategy, hence the system can adapt to different failure rates: when the failure rate is low the fraction of All-C agents is kept small (Fig. 5.3c) and thus unnecessary repairs are limited, while the fraction of All-C agents is made high when the failure rate is high. As a result of the flexible change of repair strategy, the fraction of normal agents (Fig. 5.3a) as well as available resources (Fig. 5.3b) are made stable and the difference in the failure rate is absorbed.

Fig. 5.3 SPD with strategic control with the modified payoff (available resources of the neighbor agents are added to payoff) (Oohashi and Ishida 2007). **a** Fraction of normal agents; **b** Available resources; **c** Fraction of All-C agents. Parameters are as in Table 5.1 and the initial configuration with half of All-D agents and 100 failure agents randomly chosen

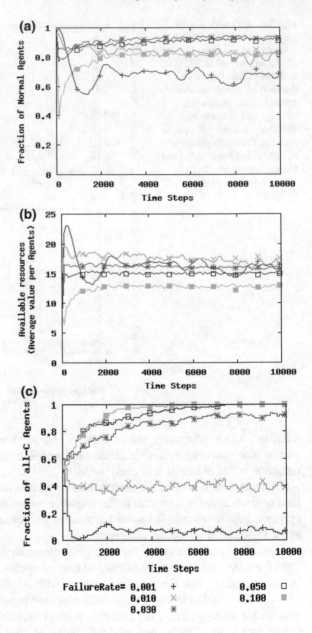

5.3.2   Adaptation to a Heterogeneous Environment

In the simulations of this heterogeneous environment and the next dynamic environment, the agents flip their action C or D by an action error rate ($\mu = 0.01$). The action error is introduced to prevent the network from becoming stuck at local minima. Further, the heterogeneous environment is implemented with a checker-board arrangement where the upper left and lower right 24 × 24 agents have a failure rate varying with relatively high values, while the upper right and lower left 24 × 24 agents have a failure rate fixed at a relatively low value of 0.001. Figure 5.4 plots the performance with varying failure rate on the horizontal axis.

Figure 5.5 (above) shows a snapshot of agent configurations at the 2000th time step when simulation is carried out with the same conditions (All-C and All-D with the systemic payoff) as those in Fig. 5.4 with the strategic repair. It can be seen that areas with higher failure rate of 0.1 (above, left and below, right) are cooperators because they require repair. On the other hand, areas with lower failure rate of 0.001 are defectors. Figure 5.5 (below), on the other hand, uses the uniform repair with a uniform repair rate of 0.5.

Fig. 5.4 SPD with strategic control with the modified payoff (available resources of the neighbor agents are added to payoff); fraction of normal agents (*above*) and fraction of All-C agents (*below*). Parameters are as in Table 5.1 and the initial configuration with half of All-D agents and 100 failure agents randomly chosen

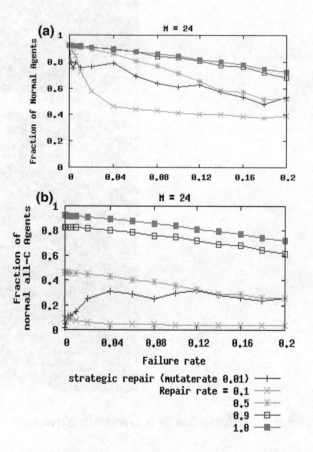

Fig. 5.5 A snapshot of agent configurations at 2000 time steps by strategic repair (*above*) and uniform probabilistic repair (*below*), when simulation is carried out with the same conditions as in Fig. 5.4. Light gray (*green*) is Normal Defector, *dark gray* (*red*) is Abnormal Defector, *white* (*yellow*) is Abnormal Cooperator and *black* (*blue*) is Normal Cooperator

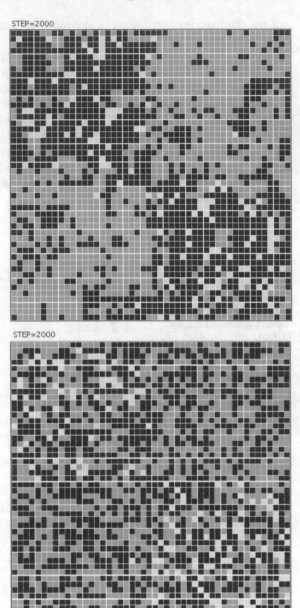

5.3.3 Adaptation to a Dynamic Environment

Since the strategic repair allows each agent to take its own action, it generally provides better performance than when the repair rate is fixed in time and is uniform among agents, particularly in a heterogeneous environment where agents fail with

Fig. 5.6 Development of performance when the failure rate changes as a triangle wave (Tokumitsu and Ishida 2008). **a** Fraction of normal agents; **b** Fraction of All-C agents; **c** Average available resources per agent; **d** Average failure rate in the network. Parameters are as in Table 5.1 and the initial configuration with the strategy of each agent and 100 failure agents randomly chosen. The failure rate is between 0.001 and 0.90

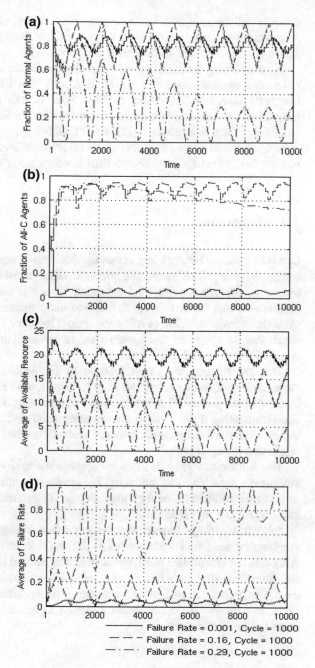

different failure rate. The strategic repair is also expected to perform well in a dynamic environment with a time-varying failure rate. In this simulation also, the action error is involved (action error rate $\mu = 0.1$).

A time-varying function is used to specify the failure rate. Agents are assumed to fail with a dynamic failure rate $\lambda(t)$ (Fig. 5.6). The failure rate oscillates with a triangular waveform as shown in Fig. 5.6d. In this simulation, the following parameters are also needed: cycle of the failure rate function (T: 10–10000) and amplitude of the failure rate function (A: 0.001–0.9).

In both spatial adaptability and temporal adaptability, a granularity of minimum volume is required for a space-and-time homogeneous area.

5.4 Discussion

In an autonomous network that leaves the decision of whether to repair neighboring agents to each selfish agent, the strategic repair exhibits adaptation to the environment. The game theoretic framework suits the autonomous and distributed decision-making context for regulation and maintenance of large-scale information systems. A major problem of using the spatial Prisoner's Dilemma in regulating the repair rate of agents is that agents tend to remain silent and stuck at the Nash equilibrium of mutual defection.

This chapter presents a new solution to this problem: that is, involving more systemic payoff incorporating not only its own resources but also all the resources in the neighborhood. With this modified payoff, agents not only have an adaptive decision-making dependent on the environmental parameters such as failure rate, but also have more favorable resource allocation when compared with a uniform regulation of repair rate.

The simulations indicated that an appropriate uniform rate could be set when parameters were identified correctly. However, it is often the case that parameters are difficult to identify, or they may change dynamically. In such cases, strategic rate control can be used.

It is worth noting, however, that defection (remaining silent without repairing the neighboring agents) may not always be unfavorable actions that should be avoided. Although the global state of all agents defecting should be avoided, remaining silent can be a wise strategy when many agents are contaminated. Whether this is true even for biological systems or whether it is specific to the network cleaning problem with cooperate and defect assigned to repair and not repair, remains an issue for further studies.

5.5 Conclusion

We have shown that a social mechanism (other than the biological one tested in
Chap. 4) of extending the payoff (systemic payoff) can also maintain the cooperators by an agent-based simulation. It was also shown that strategic repairs, even
those restricted to All-C and All-D, have adaptive capability both in space (to a
heterogeneous environment) and in time (to a dynamic environment). All-C and
All-D are virtually actions rather than spatial strategies.

In the following Chap. 6, we will see that a certain spatial strategy (kD) has even
a protective function by forming a membrane surrounding a cluster of cooperators
in a more general framework of the spatial prisoner's dilemma.

References

Hamilton, W.D.: The genetical evolution of social behaviour. I. J. Theor. Biol. **7**(1), 1–16 (1964)
Hofbauer, J., Sigmund, K.: Evolutionary game dynamics. B. Am. Math. Soc. **40**(4), 479–519
  (2003). doi:10.1090/S0273-0979-03-00988-1
Maynard Smith, J.: Group selection and kin selection. Nature **201**, 1145–1147 (1964)
Nowak, M.A.: Evolutionary Dynamics: Exploring the Equations of Life. Harvard University Press,
  (2006)
Nowak, M.A., May, R.M.: Evolutionary games and spatial chaos. Nature **359**(6398), 826–829
  (1992)
Oohashi, M., Ishida, Y.: A game theoretic approach to regulating mutual repairing in a
  self-repairing network. In: Innovative Algorithms and Techniques in Automation, Industrial
  Electronics and Telecommunications. pp. 281–286. Springer, Berlin (2007)
Taylor, P.D., Jonker, L.B.: Evolutionary stable strategies and game dynamics. Math. Biosci. **40**(1),
  145–156 (1978)
Tokumitsu, M., Ishida, Y.: Self-Repairing Network in a Dynamic Environment with a Changing
  Failure Rate. In: Novel Algorithms and Techniques in Telecommunications, Automation and
  Industrial Electronics. pp. 555–560. Springer, Berlin (2008)

Chapter 6
Protection of Cooperative Clusters by Membrane

Abstract Our spatial Prisoner's Dilemma is divided into two stages: strategy selection and action (cooperation/defection) selection (named second-order cellular automata). This renewal allows a spatiotemporal strategy that determines the player's next action based not only on the adversary's history of actions (temporal strategy) but also on neighbors' configuration of actions (spatial strategy). Several space-time parallelisms and dualisms would hold in this spatiotemporal generalization of strategy. Among them, this chapter focuses on generosity (how many defections are tolerated). A temporal strategy involving temporal generosity, such as Tit for Tat (TFT), exhibits good performance such as noise tolerance. We report that a spatial strategy with spatial generosity can maintain a cluster of cooperators by forming a membrane that protects against defectors. The condition of membrane formation can be formulated as the spatial generosity exceeding a certain threshold determined by the number of neighborhoods.

Keywords Spatial Prisoner's Dilemma · Spatiotemporal strategy · Generosity · Second-order cellular automata · Membrane formation · Protection of cooperative cluster

6.1 Introduction

In the spatial Prisoner's Dilemma, we noted that cooperators (C players) can mutually support themselves in clusters, while defectors (D players) can exploit the surrounding Cs when a single defector intrudes into the cluster of Cs. This situation can be more formally stated when applied to the evolutionary context: Evolutionarily Stable Strategy (ESS). When applied in the context of the spatial Prisoner's Dilemma, ESS is a strategy that is stable against intrusion of mutated strategies (hence a small population)

Most results of this chapter are presented in Ishida and Katsumata (2008).

© Springer International Publishing Switzerland 2015
Y. Ishida, *Self-Repair Networks*, Intelligent Systems Reference Library 101,
DOI 10.1007/978-3-319-26447-9_6

of the existing strategy (hence a cluster of large population). This chapter further extends the ESS involving geometric concepts: spatial stability of a geometric shape. That is, a geometric shape (such as a circle) is spatially stable if it rejects a certain amount of intrusion and preserves the geometric shape. As an example, we will show that a circle (or a closed loop in general) in a two-dimensional lattice can be a spatially stable shape. It functionally protects the internal cluster of cooperators inside from the intrusion of defectors outside. The approach can be also applied to the local stability of a fraction shape found in a local shape of the membrane. The ESS may be regarded as a spatially stable shape of connected clusters (*sea* as opposed to *islands*).

By restricting spatial strategies (introduced in Chaps. 4 and 5), we found that some obvious class spatial strategies would lead to a self-organization of membrane that protects the internal cluster of cooperators.

The Prisoner's Dilemma (PD) has motivated studies in many domains such as international politics and evolutionary biology since the seminal work by (Axelrod 1987, 1984; Axelrod and Hamilton 1981). The spatial Prisoner's Dilemma (SPD) devised by (Nowak and May 1992; Nowak 2006) also provides another dimension that these originally game theoretic studies can be related to the field of cellular automata (Chua 1998; Chua et al. 2002; Wolfram 1983, 2002).

Many possible mechanisms for the maintenance and protection of the cooperators' cluster have been proposed (Hauert and Doebeli 2004; McNamara et al. 2004; Szabó and Hauert 2002). In a spatiotemporal generalization of the Prisoner's Dilemma, we observed that a membrane at the perimeter of the cooperators' cluster protects the cluster from invasion by defectors where the cooperators' cluster would be invaded otherwise.

In an asymmetric interaction between cooperators and defectors, defectors can exploit cooperators if the cooperators cannot escape or cannot recognize the exploitation. Defectors can gain more as the number of exploitations increases. Usually in social interactions, cooperators recognize that they are losing each time, and escape from exploitation or become defectors. From the defectors' viewpoint, they can gain much benefit only when the adversary cooperators do not escape. In a spatial extension defectors can exploit N cooperators at once, since they need not play the game N times with a player, but the game once with N players. This is similar to spam mail; this exploitation is made possible by the Internet which allows a player to interact with so many players in one action with almost no cost.

Thus, to deal with interactions through information networks such as the Internet where one-to-many or many-to-one interactions can be done instantaneously with almost no cost, the game framework of the Prisoner's Dilemma must be extended from the conventional constraint such that a player is a unit of action, recognition, decision making, and receiving reward/penalty. (The extension seems necessary to deal with biological systems as well regarding several levels of biological units such as DNA, RNA, cells, individuals, and species as profit-pursuing players.) The extension may be theoretically possible in the following ways:

1. Scope of players on which the strategy is based;
2. Scope of players which the action will affect;
3. Scope of players whose (weighted) sum will be the payoff.

Among the three options above, this chapter will focus on the first and second ones. The scope of agents on which the strategy is based and which the action will affect is usually set to be one, however, we will extend this to many (called *neighborhood*). This extension can be considered as generalizing the conventional temporal strategy to spatiotemporal strategies. The third extension amounts to extension of players' identity to a collective identity composed of players in the neighborhood. This third extension (Ishida and Tokumitsu 2008) was already discussed as the systemic payoff (Chap. 5), but it is worth mentioning that extending players' identity will enhance and protect clusters of cooperators and bias the benefit of a set of players (neighbors) rather than a single player. Hence, with the perspective that space and time are related, the extension creates a bias toward a longer time evaluation of accumulated benefits than the conventional single player's payoff.

Section 6.2 presents our spatiotemporally generalized framework used in this chapter. Section 6.3 defines the spatial version of strategy (in contrast to the conventional temporal version) and spatial version of generosity. Section 6.4 presents the main results of the conditions for membrane formation, with simulation results.

Cooperative nodes are easier to maintain and to protect from the intrusion of defective nodes. This chapter explores the further possibility of protecting the cooperative clusters by involving spatial strategies:

Do spatial strategies have significance in protecting cooperative clusters?

Thus, the model in this chapter focuses on the spatial strategies of the spatial prisoner's dilemma, not on the self-repair. However, the model can be applied to self-repair networks as long as the payoff matrix matches the Prisoner's Dilemma.

The model assumptions of this chapter are:

- Cooperate/Defect: Each node can choose an action of cooperate or defect.
- Control: Actions (cooperate/defect) are determined based on the spatial pattern of the actions in the neighbors (spatial strategy). Once repair action is determined, it is executed deterministically. Actions are done on all the nodes simultaneously in the neighbor.
- Strategy: The strategy is updated in a constant period (strategy update cycle). In updating, the strategy of the neighbor (eight-neighbor) nodes which earned the highest payoff is copied. Payoff is measured by the total sum earned by the payoff matrix when updating the strategy.
- Network: An $L \times L$ two-dimensional square lattice with eight neighbors and with the periodic boundary condition.

6.2   Spatiotemporal Generalization of Prisoner's Dilemma

In pursuit of a mechanism that allows cooperators' clusters to be preserved, we examine a spatial version of generosity: how many defectors in the neighborhood are tolerated rather than how many previous defections were made by the adversary, in a spatiotemporally generalized context.

The action of each player is determined in two stages: the strategy is determined based on the scores in the neighborhood, then the action is determined based on the strategy and action configurations (spatial patterns) in the neighborhood. This two-stage determination of actions is similar to a second-order cellular automaton (CA) (Wolfram 2002), however, our model involves also two layered states (strategies kD and actions C/D).

Our SPD is done in the following way with N players simultaneously interacting with $(2r + 1)^2$ neighbors.

(0) **Initial arrangement**: The action and strategy of each player are determined (see Sect. 6.4 for two types of simulation).

(1) **Renewal of action**: The next action is determined by each player's strategy based on neighbors' (excluding the player itself) actions.

(2) **Score:** The score for each player is calculated by summing up all the scores received from PD with eight neighboring players and the player itself with the payoff matrix in Table 6.1, and then adding the sum to the current score of the player. (The score will be added until the strategy is renewed.)

(3) **Renewal of strategy**: After (1) and (2) are repeated q (= 1 in our simulations) times, the next strategy will be chosen from the strategy with the highest score among the neighbors and the player itself. (1) to (3) are counted as one time step.

Table 6.1 The payoff matrix of the Prisoner's Dilemma game. Payoffs are to the player in each row. A single parameter b ($1 < b < 2$) is used following Nowak-May's simulations (Nowak and May 1992)

Player	Adversary	
	ⓒ	ⓓ
ⓒ	R (1)	S (0)
ⓓ	T (b)	P (0)

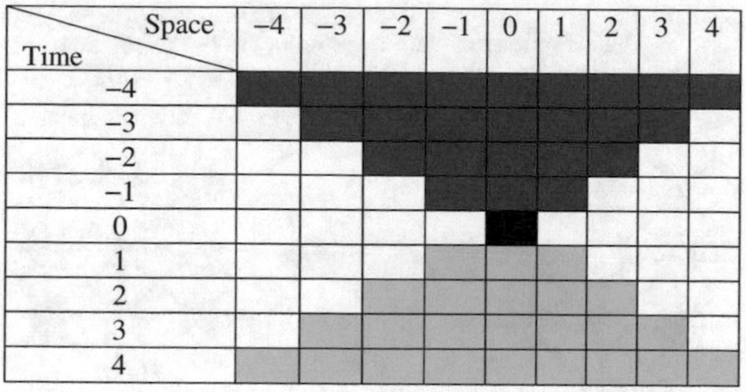

Time \ Space	-4	-3	-2	-1	0	1	2	3	4
-4									
-3									
-2									
-1									
0									
1									
2									
3									
4									

Fig. 6.1 A space-time diagram of the one-dimensional SPD when the propagation speed is 1 (neighborhood radius r = 1). The current center cell (*black*) is causally related to the cells within the cone (*triangle*). The cell can affect the cells in the cone of the future and is affected by the cells in the cone of the past

In our model, each player determines the next action by the current action of the neighbors. The strategy may be regarded as a "spatial strategy" in contrast to a "temporal strategy" in IPD.

In Fig. 6.1 (the periodic boundary in the space direction, while the discrete time is infinite in the future and past), with the temporal strategy, the current player at space 0 and time 0 determines the action based on the action history of the adversary (say, space coordinate 1) column 1 within the past cone (say, time coordinate from −1 to −3). With the spatial strategy, the current center player determines the action based on the action configuration of players −1 and 1 in row −1 within the past cone. The spatiotemporally generalized strategy is based on the action patterns in some volume of the past cone.

6.3   Spatial Strategies and Spatial Generosity

To specify a spatial strategy, actions of the eight neighbors (i.e., *Moore* neighborhood or the neighborhood radius $r = 1$) and the player itself must be specified, hence 2^9 rules are required.

For simplicity, we restrict ourselves to a "totalistic spatial strategy" that depends on the number of D (defect) actions of the neighbors, not on their positions. To represent a strategy, let l be the number of Ds of the neighboring players excluding the self.

As a typical (totalistic) spatial strategy, we define kD that takes D if $l \geq k$ and C otherwise (Fig. 6.2). For example, All-C is 9D and All-D is 0D. Note that these context independent strategies can be considered both spatial strategies and temporal strategies as extremes and bases.

This kD can be regarded as a spatial version of TFT (1D) where k seems to indicate the generosity (how many D actions in the neighborhood are tolerated).

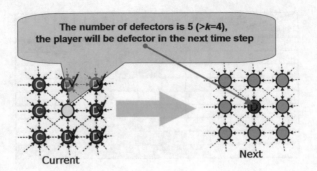

Fig. 6.2 The kD strategy as a spatial strategy. When $k = 4$ for example, the player tolerates up to three defectors in the neighborhood. However, when the number exceeds the spatial generosity k, the player will become a defector as well

6.4 Conditions for Membrane Formation

With the generalized SPD, we studied interactions between All-D and kD instead of those between All-D and All-C (as in Nowak-May's SPD). Simulations revealed that clusters of kD form a membrane of action D that protects the inner cluster of action C (note that kD can take both C and D depending on the number of Ds in the neighborhood). We observed that this membrane formation occurs (Fig. 6.3) for a certain scope of parameters k (spatial generosity), r (neighborhood radius) and b (bias for defectors in the payoff matrix in Table 6.1).

We conducted two kinds of simulations which differed in initial configuration: a random configuration and a single (symmetric centered) seed. The random configuration simulation (Fig. 6.3) is to observe membrane formation which is robust against the initial seed, and the single seed simulation (Fig. 6.4) is to investigate the conditions for membrane formation.

In both simulations, the parameter b in the payoff matrix (Table 6.1) is set to be the smallest value that allows the All-D strategy (hence defectors) to expand. That is, b is the smallest value satisfying $5b > 9$ ($r = 1$), $16b > 25$ ($r = 2$), or $33b > 49$ ($r = 3$). In the case of $r = 1$, for example, $5b$ is the maximum score that the corner of cluster D will earn and 9 is the maximum score of C which is in cluster C and located in the neighborhood of player C adjacent to corner D. Ohtsuki et al. (2006) derived an elegant formula to balance C and D by calculating fixation probabilities in a different model, however, we are unable to use it due to the form of the payoff matrix (Table 6.1).

We will focus on the condition for the membrane formation with respect to these parameters. Throughout this chapter, simulations are conducted in a square lattice with periodic boundary condition with the following parameters listed in Table 6.2.

After the membranes are formed, the following four phenomena are observed depending on the k value.

Fig. 6.3 Random configuration simulation for membrane formation in a generalized Prisoner's Dilemma. In this snapshot, All-D and kD ($k = 6$) strategies as well as C/D actions are allocated in random positions initially. The lattice size is 150×150. The time steps are 5 (*above*), 30 (*middle*), and 200 (*below*). *Black cells* are All-D; *white* and *gray cells* are respectively C and D states of kD

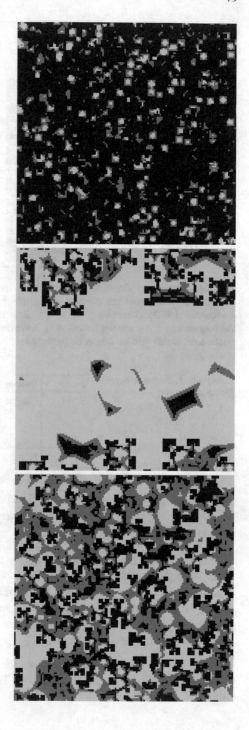

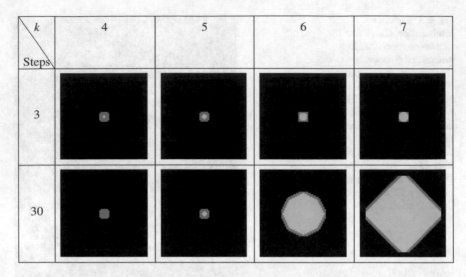

Fig. 6.4 Single seed simulations to investigate membrane formation. The kD strategy with C actions is initially set in the center with the size 4×4. The membrane is indeed formed when k exceeds 4.5 (9/2). Otherwise (as in $k = 4$), the membrane grows inside, and eventually replaces the cooperators. If k is too big (as in $k = 7$), however, the membrane will be broken and eventually cooperators inside will no longer be protected

Table 6.2 List of parameters for simulations

Name	Description	Value		
$L \times L$	Size of the space	150×150 (Fig. 6.3, Random configuration with maps, Fig. 6.4, Single seed with maps)		
T	Number of steps	200 (Fig. 6.3, Random configuration with maps)		
r	Neighborhood radius	1	2	3
b	Bias for defectors	1.81	1.57	1.485

1. k is too small: The membrane will grow toward the center of cluster kD and will corrode the C state of kD.
2. k is small: Cluster kD covered by the membrane will stay stable.
3. k is large: Cluster kD covered by the connected membrane will expand.
4. k is too large: Cluster kD covered by the broken membrane will expand and the cluster will eventually collapse.

Since we are interested in conditions for cooperators to be preserved, we focus on the conditions in cases 2 and 3 above. The spatial generosity k increases as the case proceeds downward from 1 to 4.

For a membrane formation, the spatial generosity k must exceed half of the number of players in the neighborhood:

$$k > (2r+1)^2/2.$$

Otherwise, the membrane will grow inside toward the center of cluster kD as in case 1.

For the cluster protected by the membrane to expand, the spatial generosity k must further exceed a larger threshold:

$$k > (2r+1)^2/2 + r.$$

Otherwise, the cluster does not expand although it is indeed protected by the membrane as long as it exceeds $(2r + 1)^2/2$.

In case 4, the membrane is broken if k exceeds a threshold (for example, $k = 7$ when $r = 1$ as shown in Fig. 6.4). This threshold has not been formulated yet.

This membrane formation as a phenomenon can be considered as a double-edged sword, since the membrane (kD strategy with action D) can not only protect kD players from being invaded by All-D players outside but also could replace all the players inside (kD strategy with action C). Although this double-edged sword character may be shared by the immune system, the immune system is a multi-faceted double-edged sword, since it not only protects a material identity (protein type) but also functional identities.

The double-edged sword character of the kD strategy will provide two threshold values: one for the membrane not growing toward the inside and another for the players inside the membrane not growing too fast so that the membrane will not be broken.

Membrane formation can be observed in an original SPD where the membrane is composed of cooperators following former defectors and the breached part is composed of defectors following former cooperators, when the bias for defectors is relatively high ($1.8 < b < 2$) (Nowak and May 1992) and variants (e.g. Oliveira dos Santos Soares and Martinez 2006). These phenomena indicate that the cooperators' expansion or defectors' invasion occurs in one direction for a period of time like wave propagation. In variants of the SPD with an additional third strategy such as loners in the public goods game (Szabó and Hauert 2002) and TFT (Szabo et al. 2002), it is reported that these third strategies prevent cooperators from being exploited by defectors. It is, however, not clear that these protective functions come from geometrical origins such as the membrane. It is noted that TFT also has generosity (temporal one), and a parallelism to the membrane formation involving spatial generosity can be recognized.

It is conjectured that the membrane works as a "buffer area" which can be deployed either in space (as in kD) or in time (as in TFT or loners).

6.5 Discussion

We focused on a space-time interplay through the eyes of generosity in the spa-tiotemporal Prisoner's Dilemma. However, many other results presented independently by researchers on the spatial Prisoner's Dilemma and by those on the temporal (conventional) Prisoner's Dilemma could be related in the spatiotemporal framework. As another example, we modified the payoff by adding not only the players' resources but also counting the neighbors' resources (Chap. 5). This modified payoff has an effect of favoring cooperators. Considering more collective players as a reward-receiving unit is somewhat similar to considering longer term profit in evaluating payoff.

The membrane formation and its protective function can be related to many physical phenomena such as crystal growth, and biological phenomena such as the origin of life or formation of biological units in an early stage of development.

The mechanisms and factors that determine the shape (e.g. whether the polygon is a tetragon, hexagon or octagon when the boundary is convex; fractal dimension when the boundary is a concave-convex complex) in symmetric seed simulations are left for further study.

The membrane formation in the interaction between defectors and cooperators by spatial strategies with *spatial generosity* seems robust against changes in parameters in the payoff matrix. The membrane formation can be observed even if we use payoff in the Snow Drift (chicken game) (Hauert and Doebeli 2004).

6.6 Conclusion

This chapter demonstrated a protective function of spatial strategy which is a spatially extended version of the conventional temporal strategy in a general framework of the spatial prisoner's dilemma (spatial game) not restricting ourselves to the self-repair network. We have shown that a generalized and spatially extended version of TFT can self-organize a membrane which will protect the internal cluster of cooperators under certain conditions. This extension and generalization is not new, for a generalized TFT has already been described (e.g., Watts 1999), however, this is the first report of membrane formation.

In the following Chap. 7, we will return to the self-repair network and explore the calculus of parameters exhibiting duality similar to AND-OR duality found in Boolean logic.

References

Axelrod, R.: The Evolution of Cooperation. Basic Books, New York (1984)
Axelrod, R.: The evolution of strategies in the iterated Prisoner's Dilemma. The dynamics of norms, 199–220 (1987)

Axelrod, R., Hamilton, W.D.: The evolution of cooperation. Science **211**(4489), 1390–1396 (1981)

Chua, L.O.: CNN: A Paradigm for Complexity, vol. 31. World Scientific Publishing Company, Singapore (1998)

Chua, L.O., Yoon, S., Dogaru, R.: A nonlinear dynamics perspective of Wolfram's new kind of science Part I: threshold of complexity. Int. J. Bifurcat. Chaos **12**(12), 2655–2766 (2002)

Hauert, C., Doebeli, M.: Spatial structure often inhibits the evolution of cooperation in the snowdrift game. Nature **428**(6983), 643–646 (2004)

Ishida, Y., Katsumata, Y.: A note on space-time interplay through generosity in a membrane formation with spatial Prisoner's Dilemma. In: Knowledge-Based Intelligent Information and Engineering Systems, pp. 448–455. Springer (2008)

Ishida, Y., Tokumitsu, M.: Asymmetric interactions between cooperators and defectors for controlling self-repairing. In: Knowledge-Based Intelligent Information and Engineering Systems, pp. 440–447. Springer (2008)

McNamara, J.M., Barta, Z., Houston, A.I.: Variation in behaviour promotes cooperation in the Prisoner's Dilemma game. Nature **428**(6984), 745–748 (2004)

Nowak, M.A., May, R.M.: Evolutionary games and spatial chaos. Nature **359**(6398), 826–829 (1992)

Nowak, M.A.: Evolutionary Dynamics: Exploring the Equations of Life. Harvard University Press, Cambridge (2006)

Ohtsuki, H., Hauert, C., Lieberman, E., Nowak, M.A.: A simple rule for the evolution of cooperation on graphs and social networks. Nature **441**(7092), 502–505 (2006)

Oliveira dos Santos Soares, R., Martinez, A.S.: The geometrical patterns of cooperation evolution in the spatial Prisoner's Dilemma: an intra-group model. Phys. A: Stat. Mech. Appl. **369**(2), 823–829 (2006)

Szabó, G., Hauert, C.: Phase transitions and volunteering in spatial public goods games. Phys. Rev. Lett. **89**(11), 118101 (2002)

Szabo, G., Antal, T., Szabó, P., Droz, M.: On the role of external constraints in a spatially extended evolutionary Prisoner's Dilemma game. http://www.arXiv.org.cond-mat/0205598 (2002)

Watts, D.J.: Small Worlds: The Dynamics of Networks Between Order and Randomness. Princeton University Press, Princeton (1999)

Wolfram, S.: Statistical mechanics of cellular automata. Rev. Mod. Phys. **55**(3), 601 (1983)

Wolfram, S.: A New Kind of Science, vol. 5. Wolfram Media, Champaign (2002)

Chapter 7
Duality in Logics of Self-Repair

Abstract A self-repair network consists of nodes capable of repairing other nodes where the repair success rate depends on the state (normal or abnormal) of the repairing node. This recursive structure leads to the double-edged sword of repairing, which could cause outbreaks in case the repairing causes adverse effects. The self-repair network can be equated to a probabilistic cellular automaton. Because of the distinction between repair by normal nodes and that by abnormal nodes, transition probabilities as a probabilistic cellular automaton exhibit symmetry. In this chapter, we observe that duality can serve to make many transition probabilities arranged in order. In particular, the duality that reduces to a self-dual with respect to AND-OR duality when the repair rate is close to 0 is of interest, for the parameter region is the place where computer simulation is hard. The duality also suggests that the case when the repair rate is close to 0 lies between pessimistic estimation (in repair success) of AND-repair and optimistic estimation of OR-repair.

Keywords Calculus of parameters · Duality · Mean field approximation · Probabilistic cellular automaton · Cost of autonomy · Symmetry · Boolean algebra

7.1 Introduction

As a candidate for the autonomous defense and maintenance of information systems, we consider self-repairing of a network by mutual copying (Sect. 5.2). In biological epidemics (e.g., Rhodes and Anderson 1996), spontaneous recoveries can occur and even the immune system could arise because autonomous nodes repair other nodes which leads to increased reliability, availability or longer life

---

Most results of this chapter are presented in Ishida (2010).

© Springer International Publishing Switzerland 2015
Y. Ishida, *Self-Repair Networks*, Intelligent Systems Reference Library 101,
DOI 10.1007/978-3-319-26447-9_7

time at the system level. However, distributed autonomy, which offers autonomous defense, could cause adverse effects and even complete system failure or outbreaks.

We consider the possibility of cleaning up the network by mutual copying. In information systems, repair by copying is a double-edged sword and so it is important to identify under what conditions the network can eradicate abnormal nodes from the system. We consider a probabilistic cellular automaton (p-CA) (Domany and Kinzel 1984) to model the situation where computers in a LAN perform mutual repair by copying their contents. Since the problem could lead to a critical phenomenon, repairs must be decided by considering the eradication of abnormal nodes and the network environment (Chap. 5).

Our model consists of nodes capable of repairing other connected nodes. We call the connected nodes "neighbor nodes" based on the terminology of CA. The repairing of a node may be done by overwriting its content to other nodes, since information systems depend on large-scale networks such as LAN-connected computer networks, and the grid system in electric power supply.

Once it became possible on the Internet for many selfish activities to proliferate, several utilities and protocols converged on the Nash equilibrium from which no players want to deviate (Nash 1950). Researchers focused on algorithms and computational complexity for obtaining equilibrium when selfish nodes (or agents when emphasizing autonomy) compete for resources. The cost for the Nash equilibrium relative to the optimized solution has also been discussed to measure the cost of "anarchy" (Koutsoupias and Papadimitriou 1999).

This chapter focuses on the self-maintenance task, and on self-repairs by mutual copying in particular. The double-edged sword of self-repair may be considered to be a cost of autonomy in attaining higher system reliability. However, without an autonomous distributed approach, high reliability comparable to that of biological systems will not be possible.

Section 7.2 revisits a self-repair network, and extends the model to include not only AND-repair but also OR-repair. The transition probabilities of these two types of repair are formulated in a symmetric form. Based on the transition probabilities, Sect. 7.3 presents a mean field approximation to investigate the density of abnormal nodes in the steady state.

The problem for this chapter is to find out calculation conventions focusing on the duality of AND-repair and OR-repair:

Do the formulae for the stationary values of parameters exhibit symmetry or asymmetry of duality?

To focus on the beauty of symmetries, we adopt the simplest model assumptions (similar to that of the probabilistic cellular automaton introduced in Chap. 3 except that the control is simplified; the same applies throughout Chaps. 7–10):

- Fault: A node becomes abnormal only when the repair failed.
- Repair: Not only normal nodes but also abnormal nodes can repair with a higher risk of unsuccessful repair. Repair does not require any cost.
- Control: The repair actions are executed on two neighbors simultaneously and independently with a constant rate (repair rate).
- Network: An N-node one-dimensional lattice with two neighbors and with the periodic boundary condition.

7.2   Basic Model

7.2.1   Definitions and Extensions

In a mathematical formulation, the model consists of three elements (U, T, R) where U is a set of nodes, T is a topology connecting the nodes, and R is a set of rules of the interaction among nodes. We assume that a set of nodes is a finite set with N nodes, and the topology is restricted to the one-dimensional array shown in Fig. 7.1 (which could be an n-dimensional array, complete graph, random graph, or even scale-free network) that could have S (an integer) neighbors for each node. Also, we restrict the study to the case where each node has a binary state: normal (0) or abnormal (1).

A p-CA requires probabilistic rules for interactions. The model controls the repairing of all the nodes uniformly. That is, each node tries to repair the neighbor nodes in a synchronous fashion with a probability P_r (repair rate). The repairing will be successful with the probability P_{rn} when it is done by a normal node, but with the probability P_{ra} when done by an abnormal node. The repaired nodes will be normal when all the repairing is successful (called *AND-repair*). Thus, when repairing is done by the two neighbor nodes simultaneously (called *simultaneous repair*), both of these two repairs must be successful in order for the repaired node to be normal.

In OR-repair, on the contrary, at least one repair must succeed for the simultaneous repair to succeed. Therefore, OR-repair can repair more successfully than AND-repair for the same number of repairs. The repair rate P_r controls not only the frequency of repairs but also the synchrony of repair actions: if P_r is close to 1, simultaneous repair is more likely to occur. The transition probabilities as a

Fig. 7.1 One-dimensional array with two states: normal (*0*) and abnormal (*1*)

probabilistic cellular automaton are listed in Table 7.1. In the state transition, the self-state is the center in parentheses and the two neighbor states are on the left and the right in parentheses. The self-state will be changed to the state indicated to the right of the arrow. $(001) \rightarrow 1$, for example, indicates that the normal node with right neighbor abnormal and left neighbor normal will change to abnormal with the probability stated in Table 7.1.

The transition probability satisfies several symmetries and asymmetries when they are viewed as a function of the repair parameters (P_r, P_{rn}, P_{ra}). Let $R^\wedge_{xyz}(P_r, P_{rn}, P_{ra})$ $(R^\vee_{xyz}(P_r, P_{rn}, P_{ra}))$ denote the transition probability from the configuration (xyz) to $(x1z)$ as a function of the parameters when the AND-repair (OR-repair) is applied where $x, y, z \in \{0, 1\}$. For example, $R^\wedge_{000}(P_r, P_{rn}, P_{ra}) = P_r(1 - P_{rn})(2 - P_r + P_r P_{rn})$. The bar as in $\bar{p}$ indicates negation (flipping probability p to $1-p$). Then, the following properties can be observed.

Duality in P_{rn}, P_{ra} exchange:

Dual: $\quad R^\wedge_{1y1}(P_r, P_{ra}, P_{rn}) = R^\wedge_{0y0}(P_r, P_{rn}, P_{ra})$, $R^\wedge_{0y0}(P_r, P_{ra}, P_{rn}) = R^\wedge_{1y1}(P_r, P_{rn}, P_{ra})$

Self-dual: $\quad R^\wedge_{0y1}(P_r, P_{ra}, P_{rn}) = R^\wedge_{0y1}(P_r, P_{rn}, P_{ra})$, $R^\wedge_{1y0}(P_r, P_{ra}, P_{rn}) = R^\wedge_{1y0}(P_r, P_{rn}, P_{ra})$

Non-sensitive: $\quad R^\wedge_{0y0}(P_r, P_{rn}, P_{ra}) = R^\wedge_{0y0}(P_r, P_{rn}, -)$, $R^\wedge_{1y1}(P_r, P_{rn}, P_{ra}) = R^\wedge_{1y1}(P_r, -, P_{ra})$

Degenerate: $\quad R^\wedge_{0y1}(P_r, P_{rn}, P_{rn}) = R^\wedge_{0y0}(P_r, P_{rn}, P_{ra})$, $R^\wedge_{0y1}(P_r, P_{ra}, P_{ra}) = R^\wedge_{0y1}(P_r, P_{rn}, P_{ra})$

Duality in P_{rn}, P_{ra} flip:

$$R^\wedge_{xyz}(P_r, P_{rn}, P_{ra}) = \overline{\left(R^\vee_{x\bar{y}z}(P_r, \overline{P_{rn}}, \overline{P_{ra}})\right)},$$
$$R^\vee_{xyz}(P_r, P_{rn}, P_{ra}) = \overline{\left(R^\wedge_{x\bar{y}z}(P_r, \overline{P_{rn}}, \overline{P_{ra}})\right)}.$$

When expressed in a form similar to *de Morgan's* theorem:

$$\overline{R^\wedge_{xyz}(P_r, P_{rn}, P_{ra})} = \left(R^\vee_{x\bar{y}z}(P_r, \overline{P_{rn}}, \overline{P_{ra}})\right),$$
$$\overline{R^\vee_{xyz}(P_r, P_{rn}, P_{ra})} = \left(R^\wedge_{x\bar{y}z}(P_r, \overline{P_{rn}}, \overline{P_{ra}})\right).$$

The involvement of probability in logic reveals a continuous structure of the logical operators AND and OR. Indeed, we can observe when the distinction

Table 7.1 Rules for state transition in the probabilistic cellular automaton

State transition	Transition probability (AND-repair)	Transition probability (OR-repair)
$(000) \rightarrow 1$	$P_r(1 - P_{rn})(2 - P_r + P_r P_{rn})$	$P_r(1 - P_{rn})(2 - P_r - P_r P_{rn})$
$(001) \rightarrow 1$	$P_r^2(1 - P_{rn}P_{ra}) + P_r(1 - P_r)((1 - P_{rn}) + (1 - P_{ra}))$	$P_r(2 - P_r - P_{rn} - P_{ra} + P_r P_{rn} P_{ra})$
$(101) \rightarrow 1$	$P_r(1 - P_{ra})(2 - P_r + P_r P_{ra})$	$P_r(1 - P_{ra})(2 - P_r - P_r P_{ra})$
$(010) \rightarrow 1$	$1 - P_r P_{rn}(2(1 - P_r) + P_r P_{rn})$	$(1 - P_r P_{rn})^2$
$(011) \rightarrow 1$	$1 - P_r((P_{rn} + P_{ra})(1 - P_r) + P_r P_{ra} P_{rn})$	$1 - P_r(P_{rn} + P_{ra} - P_r P_{rn} P_{ra})$
$(111) \rightarrow 1$	$1 - P_r P_{ra}(2(1 - P_r) + P_r P_{ra})$	$(1 - P_r P_{ra})^2$

between AND and OR disappears. When Pr is close to 0 ($P_r \cong 0$), we will write as (0) instead of Pr. Thus, $R^{\vee}{}_{xyz}(P_r, P_{rn}, P_{ra}) = R^{\vee}{}_{xyz}((0), P_{rn}, P_{ra})$ when Pr is close to 0. It is also noted that the repair rate P_r can be an indicator of synchrony in repairing: the repair is synchronous when it is close to 1, and asynchronous when close to 0 (measured by the probability of simultaneous repair).

Although it is tempting to formulate the self-repair network based on the formulation (starting from the algebra satisfying the above), lack of space precludes us from doing so here.

Since AND-repair and OR-repair match when the second-order terms of P_r are neglected (Table 7.2), we will use $R_{xyz}((0), P_{rn}, P_{ra})$ for both of them:

$$R^{\vee}{}_{xyz}((0), P_{rn}, P_{ra}) = R^{\wedge}{}_{xyz}((0), P_{rn}, P_{ra}) \ (\equiv R_{xyz}((0), P_{rn}, P_{ra}))$$

The above expression in a form similar to *de Morgan's* theorem, but self-dual with no twin form, can be a simpler form when Pr is close to 0:

$$\overline{R_{xyz}((0), P_{rn}, P_{ra})} = R_{x\bar{y}z}((0), \overline{P_{rn}}, \overline{P_{ra}})$$

Using $R_{xyz}((0), P_{rn}, P_{ra})$ which appears both in AND-repair and OR-repair, they can be formulated in a simple form (Tables 7.2 and 7.3).

It is first noted that $R_{xyz}((0), P_{rn}, P_{ra})$ has symmetry in flipping P_{rn} and P_{ra}:

$$R_{xyz}((0), P_{rn}, P_{ra}) = \overline{\left(R_{x\bar{y}z}((0), \overline{P_{rn}}, \overline{P_{ra}})\right)}$$

Further, the following inequalities hold:

$$R_{x0z}((0), P_{rn}, P_{ra}) > R^{\wedge}{}_{x0z}((1), P_{rn}, P_{ra}) > R^{\vee}{}_{x0z}((1), P_{rn}, P_{ra}),$$
$$R^{\wedge}{}_{x1z}((1), P_{rn}, P_{ra}) > R^{\vee}{}_{x1z}((1), P_{rn}, P_{ra}) > R_{x1z}((0), P_{rn}, P_{ra}).$$

From this Table 7.3, the following inequalities are observed:

$$R_{x0z}((0), P_{rn}, P_{ra}) > R^{\wedge}{}_{x0z}(P_r, P_{rn}, P_{ra}) > R^{\vee}{}_{x0z}(P_r, P_{rn}, P_{ra}),$$
$$R^{\wedge}{}_{x1z}(P_r, P_{rn}, P_{ra}) > R^{\vee}{}_{x1z}(P_r, P_{rn}, P_{ra}) > R_{x1z}((0), P_{rn}, P_{ra}).$$

7.2.2 Related Models

The Domany-Kinzel (DK) model (Domany and Kinzel 1984) is a one-dimensional two-state and totalistic p-CA in which the interaction timing is specific. The interaction is done in an alternated synchronous fashion: the origin cell with state 1 is numbered as 0. The numbering proceeds $\{1, 2, \ldots\}$ to the right, and $\{-1, -2, \ldots\}$ to the left. At the Nth step the even numbered cells will act on the odd numbered

Table 7.2 Rules for state transition in the probabilistic cellular automaton (first-order approximation)

State transition	Transition probability (common to both AND-and OR-repair) $P_r \cong 0$ (the second-order terms of P_r are neglected)	Transition probability (AND-repair) $P_r \cong 1$ (the second-order terms of $\overline{P_r}$ are neglected)	Transition probability (OR-repair) $P_r \cong 1$ (the second-order terms of $\overline{P_r}$ are neglected)
(000) $\to$ 1	$2P_r\overline{P_m}$	$R_{xyz}((0), P_m, P_{ra}) - (\overline{P_m})^2 (P_r - \overline{P_r})$	$R_{xyz}((0), P_m, P_{ra}) - (\overline{P_m})(1+P_m)(P_r - \overline{P_r})$
(001) $\to$ 1	$P_r(\overline{P_m} + \overline{P_{ra}})$	$R_{xyz}((0), P_m, P_{ra}) - (\overline{P_m})(P_{ra})(P_r - \overline{P_r})$	$R_{xyz}((0), P_m, P_{ra}) - (1 - P_m P_{ra})(P_r - \overline{P_r})$
(101) $\to$ 1	$2P_r\overline{P_{ra}}$	$R_{xyz}((0), P_m, P_{ra}) - (\overline{P_{ra}})^2 (P_r - \overline{P_r})$	$R_{xyz}((0), P_m, P_{ra}) - (\overline{P_{ra}})(1+P_{ra})(P_r - \overline{P_r})$
(010) $\to$ 1	$1 - 2P_r P_m$	$R_{xyz}((0), P_m, P_{ra}) + P_m(1+\overline{P_m})(P_r - \overline{P_r})$	$R_{xyz}((0), P_m, P_{ra}) + P_m^2 (P_r - \overline{P_r})$
(011) $\to$ 1	$1 - P_r(P_m + P_{ra})$	$R_{xyz}((0), P_m, P_{ra}) + (P_{ra} + P_m \overline{P_{ra}})(P_r - \overline{P_r})$	$R_{xyz}((0), P_m, P_{ra}) + P_m P_{ra} (P_r - \overline{P_r})$
(111) $\to$ 1	$1 - 2P_r P_{ra}$	$R_{xyz}((0), P_m, P_{ra}) + P_{ra}(1+\overline{P_{ra}})(P_r - \overline{P_r})$	$R_{xyz}((0), P_m, P_{ra}) + P_{ra}^2 (P_r - \overline{P_r})$

Table 7.3 Rules for state transition in the probabilistic cellular automaton

State transition	Transition probability $R_{xyz}((0), P_{rm}, P_{ra})$	Transition probability (AND-repair) $R_{xyz}^{\wedge}(P_r, P_{rm}, P_{ra})$	Transition probability (OR-repair) $R_{xyz}^{\vee}(P_r, P_{rm}, P_{ra})$
$(000) \to 1$	$2P_r\overline{P_{rm}}$	$R_{xyz}((0), P_{rm}, P_{ra}) - P_r^2(\overline{P_{rm}})^2$	$R_{xyz}((0), P_{rm}, P_{ra}) - P_r^2(\overline{P_{rm}})(1 + P_{rm})$
$(001) \to 1$	$P_r(\overline{P_{rm}} + \overline{P_{ra}})$	$R_{xyz}((0), P_{rm}, P_{ra}) - P_r^2(\overline{P_{rm}})(\overline{P_{ra}})$	$R_{xyz}((0), P_{rm}, P_{ra}) - P_r^2(1 - P_{rm}P_{ra})$
$(101) \to 1$	$2P_r\overline{P_{ra}}$	$R_{xyz}((0), P_{rm}, P_{ra}) - P_r^2(\overline{P_{ra}})^2$	$R_{xyz}((0), P_{rm}, P_{ra}) - P_r^2(\overline{P_{ra}})(1 + P_{ra})$
$(010) \to 1$	$1 - 2P_rP_{rm}$	$R_{xyz}((0), P_{rm}, P_{ra}) + P_r^2P_{rm}(1 + \overline{P_{rm}})$	$R_{xyz}((0), P_{rm}, P_{ra}) + P_r^2P_{rm}^2$
$(011) \to 1$	$1 - P_r(P_{rm} + P_{ra})$	$R_{xyz}((0), P_{rm}, P_{ra}) + P_r^2(P_{ra} + P_{rm}\overline{P_{ra}})$	$R_{xyz}((0), P_{rm}, P_{ra}) + P_r^2P_{rm}P_{ra}$
$(111) \to 1$	$1 - 2P_rP_{ra}$	$R_{xyz}((0), P_{rm}, P_{ra}) + P_r^2P_{ra}(1 + \overline{P_{ra}})$	$R_{xyz}((0), P_{rm}, P_{ra}) + P_r^2P_{ra}^2$

Table 7.4 Rules for the DK model where p_1 and p_2 are two parameters for the DK model and symbol * is a wildcard

State transition	Transition probability
$(0*0) \to 0$	1
$(0*1) \to 1$	p_1
$(1*1) \to 1$	p_2

cells and the odd numbered cells will act at the next step. The neighbors are the two cells adjacent to oneself without self-interaction. The interaction rule is as shown in Table 7.4.

The self-repair network with AND-repair can be equated to the DK model (Chap. 3 and Ishida 2005a, b) when $P_r = 1$ (i.e. nodes always repair) with the parameters $p_1 = 1 - P_{ra}$, $p_2 = 1 - P_{ra}^2$, i.e., the case of the directed bond percolation.

7.3 Steady-State Analysis

7.3.1 Mean Field Approximation

Let ρ_1 ($\rho_0 = 1 - \rho_1$) be a mean field approximation of the density of abnormal nodes (normal nodes). The dynamics of ρ_1 can be described by the following equation by letting a, b, c, d, e and f denote the transition probabilities $(000) \to 1$, $(001) \to 1$, $(101) \to 1$, $(010) \to 0$, $(011) \to 0$, and $(111) \to 0$, respectively.

$$\frac{d\rho_1}{dt} = a\rho_0^3 + 2b\rho_0^2\rho_1 + c\rho_0\rho_1^2 - d\rho_0^2\rho_1 - 2e\rho_0\rho_1^2 - f\rho_1^3$$

By eliminating ρ_0, we obtain the following equation with only ρ_1:

$$\frac{d\rho_1}{dt} = A\rho_1^3 + B\rho_1^2 + C\rho_1 + D$$

where A, B, C and D are constants determined by the three parameters of the self-repair network as follows:

$$A = -(a+c+d+f)+2(b+e),$$
$$B = (3a+c+2d)-2(2b+e),$$
$$C = -(3a+d)+2b,$$
$$D = a.$$

As in the transition probability (Table 7.5), a first-order approximation again makes no difference between AND-repair and OR-repair.

The mean field approximation of the ratio of abnormal nodes ρ_1 may be obtained by solving a second-order algebraic equation; however, the root can be a complicated form in general. Nevertheless, some of them can be simple in specific cases.

First, when $P_{rm} = 1$ both AND-repair and OR-repair have a simple solution. For abnormal nodes eradicated in AND-repair (Chap. 3), the (abnormal nodes) eradication condition $\frac{P_{ra}}{P_r} \geq \frac{1}{2}$ must be satisfied, since the time derivative $\frac{d\rho_1}{dt}$ must be negative. Otherwise, the density of abnormal nodes converges on $\frac{P_r - 2P_{ra}}{P_r(1-P_{ra})^2}$.

In OR-repair, for example, the density of abnormal nodes converges on 0. This means that OR-repair can eradicate all the abnormal nodes at $P_{rm} = 1$ regardless of the values of P_r and P_{ra}.

When $P_{rm} = P_{ra}$, the fixed point of the equation is $\frac{(1-P_m)(2-P_r+P_rP_m)}{2-P_r}$ for AND-repair, and $\frac{(1-P_m)(2-P_r-P_rP_m)}{2-P_r}$ for OR-repair. Since they are stable points, the density of abnormal nodes converges on the points.

When the first-order approximation of P_r is adopted, the fixed point of the equation is $\frac{1-P_m}{1-P_m+P_{ra}}$, which is stable for both AND-repair and OR-repair, hence the density of abnormal nodes converges on the point.

7.3.2 Synchrony or Asynchrony

In the self-repair network, the chances for repair are given equally and synchronously to all nodes. However, whether the action of repair takes place or not is a probabilistic event. (Further, even when the repair action takes place, whether the repair is successful or not is another probabilistic event.) Thus, although the synchrony of repair actions is not explicitly controlled, it is controlled through the repair rate: the closer P_r is to 1, the more synchronous repair actions are.

Because we have the following inequality, if there are many abnormal (1) nodes, P_r closer to 0 (but not 0), hence more asynchrony, is favored (to eradicate abnormal

Table 7.5 Coefficients of the equation expressed by parameters of the self-repair network

Constant	Constant expressed by parameters (AND-repair)	Constant expressed by parameters (OR-repair)
A	0	0
B	$-P_r^2(P_m - P_{ra})^2$	$P_r^2(P_m - P_{ra})^2$
C	$2P_r(P_m - P_{ra} - 1) + P_r^2\{-2(1 - P_m)(P_m - P_{ra}) + 1\}$	$2P_r(P_m - P_{ra} - 1) + P_r^2\{-2P_m(P_m - P_{ra}) + 1\}$
D	$2P_r(1 - P_m) - P_r^2(1 - P_m)^2$	$2P_r(1 - P_m) - P_r^2(1 - P_m^2)$

nodes). If there are more normal (0) nodes, P_r closer to 1, hence more synchrony, is favored. Further, if we can choose AND-repair or OR-repair, OR-repair is favored.

$$R_{x0z}((0), P_{rn}, P_{ra}) > R^\wedge_{x0z}((1), P_{rn}, P_{ra}) > R^\vee_{x0z}((1), P_{rn}, P_{ra})$$

$$R^\wedge_{x1z}((1), P_{rn}, P_{ra}) > R^\vee_{x1z}((1), P_{rn}, P_{ra}) > R_{x1z}((0), P_{rn}, P_{ra})0$$

Since we have the following inequalities, OR-repair is always favored if we were to choose AND-repair or OR-repair. However, P_r closer to 0 is favored only when there are relatively many abnormal nodes.

$$R_{x0z}((0), P_{rn}, P_{ra}) > R^\wedge_{x0z}(P_r, P_{rn}, P_{ra}) > R^\vee_{x0z}(P_r, P_{rn}, P_{ra})$$

$$R^\wedge_{x1z}(P_r, P_{rn}, P_{ra}) > R^\vee_{x1z}(P_r, P_{rn}, P_{ra}) > R_{x1z}((0), P_{rn}, P_{ra})$$

7.4 Summary and Discussion

Our interest in the self-repair network is also motivated by von Neumann's probabilistic logic and we attempted to explore the conjecture that: "The ability of a natural organism to survive in spite of a high incidence of error (which our artificial automata are incapable of) probably requires a very high flexibility and ability of the automaton to watch itself and reorganize itself. And this probably requires a very considerable autonomy of parts" (Von Neumann and Burks 1966).

Although we focused on a self-repair task as an example, it should be stressed that the structure becomes recursive such that the task depends upon the node that is doing the task. Indeed, for a case of repair, the repair success rate depends upon whether the repairing node is normal (0) or abnormal (1). In a conventional logic such as Boolean logic (Boole 1854), true (0) or false (0) of a logical variable will not in general affect the truth value of other logical variables.

This recursive structure in a logic leads to a complicated form in state transition probability, which depends upon not only the repair rate but also the repair success rate (by abnormal nodes as well as normal ones). Also, this recursive structure creates a double-edged sword effect of the repairing actions.

Nevertheless, several symmetries and asymmetries are identified in terms of several operations such as an exchange of *twin* probabilities: P_{rn} and P_{ra}; and logical operators AND, OR and negation (flipping probability p to $1 - p$). These symmetries (and asymmetries) further result in duality when expressed in a form similar to logic. Although a duality similar to Boolean logic (Boole 1854) holds, the attempt here shares a *self-reference* with Brown logic (Spencer-Brown 1969; Varela 1979).

7.5   Conclusion

Symmetry requires a mapping to define it mathematically. An object has symmetry if it is indistinguishable from the one mapped. Duality is also a special type of symmetry, for the dual of the dual must be the original. This book suggests several concepts that can be dual: a chain repair to a chain failure (cascade failure); the double-edged sword of repairing to the dilemma nature of being repaired (some components must be repaired from outside and at the same time must be disconnected from others for containment); repairing to infection; and so on. Among others, this chapter showed that the transition probability satisfies duality when the repair rate P_r is close to 0 in exchanging AND and OR operators in the repair scheme. In fact, this duality turns out to be self-dual where there are no twin forms similar to *de Morgan's* theorem, hence no distinction between AND-repair and OR-repair. This duality can shed light on probabilistically reduced-size modeling and simulation.

In the following Chap. 8, we will see that repairing and infection are not dual but asymmetric.

References

Boole, G.: The Laws of Thought, vol. 2. The Open Court Publishing Company (1854)

Domany, E., Kinzel, W.: Equivalence of cellular automata to Ising models and directed percolation. Phys. Rev. Lett. **53**(4), 311–314 (1984)

Ishida, Y., Mori, T.: A network self-repair by spatial strategies in spatial prisoner's dilemma. In: Knowledge-Based Intelligent Information and Engineering Systems, pp. 79–85, Springer (2005a)

Ishida, Y., Mori, T.: Spatial strategies in a generalized spatial prisoner's dilemma. Artif. Life Robot. **9**(3), 139–143 (2005b)

Ishida, Y.: A note on symmetry in logic of self-repair: the case of a self-repair network. In: Knowledge-Based and Intelligent Information and Engineering Systems, pp. 652–659. Springer, Berlin (2010)

Koutsoupias, E., Papadimitriou, C.: Worst case equilibria. In: 16th Annual Symposium on Theoretical Aspects of Computer Science—Stacs'99, vol. 1563, pp. 404–413 (1999)

Nash, J.F.: The bargaining problem. Econometrica: J. Economet. Soc. **18**(2), 155–162 (1950)

Rhodes, C.J., Anderson, R.M.: Dynamics in a lattice epidemic model. Phys. Lett. A **210**(3), 183–188 (1996). doi:10.1016/S0375-9601(96)80007-7

Spencer-Brown, G.: Laws of Form. Allen and Unwin, London (1969)

Varela, F.J.: Principles of Biological Autonomy. Elsevier, North Holland (1979)

Von Neumann, J.V., Burks, A.W.: Theory of Self-Reproducing Automata. University of Illinois Press, Urbana and London (1966)

Chapter 8
Asymmetry Between Repair and Infection in Self-Repair Networks

Abstract A self-repair network is a model consisting of autonomous nodes capable of repairing connected nodes. Self-repairing by mutual repair involves the double-edged sword problem where repairing could cause adverse effects when done by abnormal nodes. Although self-repair in a single node (as opposed to mutual repair between two nodes) is not susceptible to this problem because an abnormal node always repairs itself, the possibility of normal nodes repairing abnormal nodes is lost. This chapter compares these two types of repair: self and mutual for a self-repair network with infections. With this extended model, abnormal nodes can spread not only by repair failures but also by infections. Although the repair (normal state spreading) can compensate for the infection (abnormal state spreading), we discuss why repairing is asymmetric to the infection.

Keywords Infection · Asymmetry · Probabilistic cellular automaton · Distributed autonomy · Mutual repair · Self-repair

8.1 Introduction

Recent progress in network technology has made possible not only huge computer networks but also several innovative systems based on them such as cloud computing, grid computing (Foster and Kesselman 2001; Foster and Kesselman 2003; Foster et al. 1998) and parasitic computing (Barabási et al. 2001). However, large-scale information systems involve such risks as large-scale malfunctions and even outbreaks of infections. When systems become too large to be dealt with by only a central authority, distributed autonomy will be essential. Although distributed autonomy incurs the risk of malicious agents other than machine failures, it is useful not only for control and management purposes but also for robustness. Further, computer viruses exploit networked computers as infection paths, and

---

Most results of this chapter are presented in Ishida and Tanabe (2010).

© Springer International Publishing Switzerland 2015
Y. Ishida, *Self-Repair Networks*, Intelligent Systems Reference Library 101,
DOI 10.1007/978-3-319-26447-9_8

hence anti-virus programs must be installed on the computers in a distributed autonomous fashion rather than a central management fashion.

Regarding distributed autonomy, we examined a network cleaning problem and proposed a self-repair network model (Chap. 3) that can be examined using knowledge of probabilistic cellular automata (Domany and Kinzel 1984). This chapter reports on the model involving infection in addition to repairing (whose effect seems asymmetric to the infection).

Section 8.2 revisits the self-repair network and extends it by involving infections as well as repair. Section 8.3 examines the model with a steady-state analysis and a computer simulation. Section 8.4 discusses the implications of the model and simulation considering the current situation of computer networks.

The problem for this chapter is to evaluate the asymmetry of two processes, i.e., infection and repair:

> Are the two fundamental processes of infection and repair symmetric or asymmetric in self-repair networks?

To keep the analyses and simulations simple, we adopt the simplest model assumptions (similar to that of the probabilistic cellular automaton introduced in Chap. 3 except the control is simplified; the same applies throughout Chaps. 7–10):

- Fault: A node becomes abnormal not only when the repair failed but also independently with a constant rate (failure rate).
- Repair: Not only normal nodes but also abnormal nodes can repair with a higher risk of unsuccessful repair. Repair does not require any cost.
- Control: The repair actions are executed on two neighbors simultaneously and independently with a constant rate (repair rate).
- Network: An N-node one-dimensional lattice with two neighbors and with the periodic boundary condition.

8.2 Basic Model

8.2.1 Definitions and Extensions

The self-repair network consists of autonomous nodes capable of repairing neighbor nodes (i.e. connected nodes) by copying their contents. Each node has a binary state: normal (0) or abnormal (1). Each node repairs the neighbor nodes with a probability P_r (called *repair rate*). The repair will succeed with a probability P_{rn} (called *repair success rate by normal nodes*) when it is done by a normal node, but with a probability P_{ra} (called *repair success rate by abnormal nodes*) when done by an abnormal node.

Repair may be divided into two types depending on the target of repair: *self-repair* (Fig. 8.1a) targets the repairing node itself; *mutual repair* (Fig. 8.1b) targets the node connected to the repairing node. In the self-repair by each node, there is no interaction among nodes and hence the transition is trivial (Table 8.1). In the state transition, the self-state is shown in parentheses. The self-state will be changed to the state indicated to the right of the arrow.

Mutual repair can be further subdivided into two types: AND-repair and OR-repair. In AND-repair, all the repairs must be successful when the repairs are done by multiple nodes simultaneously, while OR-repair requires at least one repair to be successful out of multiple repairs done simultaneously. We consider only AND-repair here, for OR-repair can eradicate abnormal nodes when repair done by normal nodes always succeeds ($P_{rn} = 1$).

In a one-dimensional array with two adjacent neighbor nodes (Fig. 8.1c), the probabilities for each state transition are listed in Table 8.2. In the state transition,

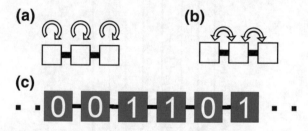

Fig. 8.1 a Self-repair. **b** Mutual repair. **c** One-dimensional array with two states: normal (*0*) and abnormal (*1*). Each node is connected to two nodes, one on either side (neighbor nodes)

Table 8.1 Transition probability in each state transition (self-repair)

State transition	Transition probability
(0)→1	$P_r(1 - P_{rn})$
(1)→1	$P_r(1 - P_{ra}) + (1 - P_r)$

Table 8.2 Transition probability in each state transition

State transition	Transition probability (mutual AND-repair)	Transition probability (infection)
(000) → 1	$P_r(1 - P_{rn})(2 - P_r + P_r P_{rn})$	0
(001) → 1	$P_r^2(1 - P_{rn}P_{ra}) + P_r(1 - P_r)((1 - P_{rn}) + (1 - P_{ra}))$	P_i
(101) → 1	$P_r(1 - P_{ra})(2 - P_r + P_r P_{ra})$	$P_i(2 - P_i)$
(010) → 1	$1 - P_r P_{rn}(2(1 - P_r) + P_r P_{rn})$	1
(011) → 1	$1 - P_r((P_{rn} + P_{ra})(1 - P_r) + P_r P_{ra}P_{rn})$	1
(111) → 1	$1 - P_r P_{ra}(2(1 - P_r) + P_r P_{ra})$	1

Table 8.3 Transition probability in each state transition (infection before repair)

State transition	Transition probability (mutual AND-repair)	Transition probability (self-repair)
(000) → 1	$P_r(1-P_m)(2-P_r+P_rP_m)$	$P_r(1-P_m)$
(001) → 1	$P_r^2(1-P_mP_{ra})+P_r(1-P_r)((1-P_m)+(1-P_{ra}))+P_i(1-P_r)^2$	$(1-P_i)P_r(1-P_m)+P_i(1-P_rP_{ra})$
(101) → 1	$P_r(1-P_r)(2-P_r+P_rP_{ra})+P_i(2-P_i)(1-P_r)^2$	$(1-P_i(2-P_i))P_r(1-P_m)+P_i(2-P_i)(1-P_rP_{ra})$
(010) → 1	$1-P_rP_m(2(1-P_r)+P_rP_m)$	$1-P_rP_{ra}$
(011) → 1	$1-P_r((P_m+P_{ra})(1-P_r)+P_rP_{ra}P_m)$	$1-P_rP_{ra}$
(111) → 1	$1-P_rP_{ra}(2(1-P_r)+P_rP_{ra})$	$1-P_rP_{ra}$

the self-state is the center in parentheses and the two neighbor states are on the left and the right in parentheses. The self-state will be changed to the state indicated to the right of the arrow. $(001)\rightarrow 1$, for example, indicates that the normal node with right neighbor abnormal and left neighbor normal will change to abnormal with the probability stated in Table 8.2.

When infection is involved in a one-dimensional array with a probability P_i (called *infection rate*), the probability of each state transition is listed in the right-most column of Table 8.2. Infection occurs only when at least one infected node exists in the neighborhood.

8.2.2   Repair Coupled with Infection

When repair is coupled with infection, the order of the repair and infection counts, that is, whether infection is after repair or before repair leads to different results. We focus on the case in which infection occurs before repair (Table 8.3), for simulations indicated that the case of repair before infection makes only a little difference.

8.3   Analyses and Simulations

8.3.1   Steady-State Analysis

Under the approximation that the probability of the state of a node being abnormal is constant and equated with a density ρ_1 ($\rho_0 = 1 - \rho_1$) of abnormal nodes (mean field approximation and steady state), the following differential equation is obtained:

$$\frac{d\rho_1}{dt} = A\rho_1^3 + B\rho_1^2 + C\rho_1 + D,$$

where the coefficients A, B, C and D are constants determined by the parameters of the self-repair network (Table 8.4).

Three roots are obtained in the third-order algebraic equation given by the steady state, that is, the roots of $A\rho_1^3 + B\rho_1^2 + C\rho_1 + D = 0$. For simplicity, we limit ourselves to the case that the repair by normal nodes always succeeds ($P_{rn} = 1$). These three roots are fixed points of the differential equation above, and we can obtain the density of abnormal nodes in the steady state determined by the stable point corresponding to the root. Figures 8.2 and 8.3 plot the density obtained by a numerical study of the root of the above algebraic equation.

Although mutual repair involves both a single repair (repair from one node) and a simultaneous repair (repair from two nodes, one on either side), self-repair

Table 8.4 Coefficients of the equation expressed by parameters of the self-repair network (infection before repair)

Constant	Constants expressed by parameters (mutual AND-repair)	Constants expressed by parameters (self-repair)
A	$P_i^2(1 - P_r)^2$	$P_i^2\{(1 - P_r P_{ra}) - P_r(1 - P_m)\}$
B	$-P_r^2(P_{rm} - P_{ra})^2 - P_i(2 + P_i)(1 - P_r)^2$	$-P_i(2 + P_i)\{(1 - P_r P_{ra}) - P_r(1 - P_m)\}$
C	$-2P_r(1 - P_m)(P_r(P_{rm} - P_{ra}) + 1) +$ $P_r(P_r - 2P_{ra}) + 2P_i(1 - P_r)^2$	$-P_r(1 + 2P_i)(1 - P_{rm} + P_{ra}) + 2P_i$
D	$P_r(1 - P_{rm})(2 - P_r + P_r P_{rm})$	$P_r(1 - P_m)$

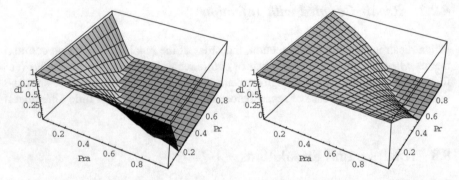

Fig. 8.2 Density of abnormal nodes (d_l) plotted for an infection rate of 0.1 (*left*) and 0.5 (*right*) as a function of the repair success rate by abnormal nodes (P_{ra}) and the repair rate (P_r) in self repair

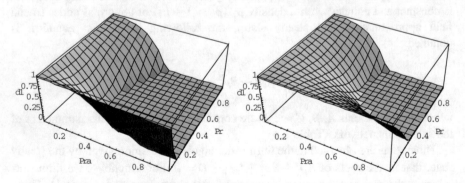

Fig. 8.3 Density of abnormal nodes (d_l) plotted for an infection rate of 0.1 (*left*) and 0.5 (*right*) as a function of the repair success rate by abnormal nodes (P_{ra}) and the repair rate (P_r) in mutual repair

includes only the single repair (repair by itself). Under the mean field approximation, averaged repair success rates for the following two cases are:

$\{(1 - \rho_1) + P_{ra}\rho_1\}^2$ simultaneous repair of a node by two nodes, one on either side;

$(1 - \rho_1) + P_{ra}\rho_1$ single repair of a node by one node on one side

In the mutual repair with AND-repair scheme, both simultaneous repair and single repair occur (although in the self-repair, only single repair occurs). Because the repair success rate of single repair is greater than that of simultaneous repair, single repair is favored. In mutual repair, since single repair occurs (with a rate $2P_r(1 - P_r)$) more often than simultaneous repair (with a rate P_r^2) when $P_r < 2/3$, mutual repair can utilize repairs more successfully if the same amount of repairs is allowed.

8.3.2   Simulation Results

Simulations are conducted with the following parameters (Table 8.5) under the conditions that the repair by normal nodes always succeeds ($P_{rn} = 1$) and infections occur before repairing.

Figure 8.4 shows a phase diagram plotted by the simulations when the infection rate is changed. Each plot indicates when all the abnormal nodes can be eradicated. The curves in each plot separate the area into two parameter regions: lower-left (including the origin) and upper-right (including upper-right $P_r = 1$ and $P_{ra} = 1$). The lower-left region is the *active phase* where some abnormal nodes remain in the network, and the upper-right region is the *frozen phase* where all the nodes become normal.

As the infection rate increases from 0.0 to 1.0 (only 0.1 and 0.5 shown), these curves are dragged to the upper-right, for the frozen region shrinks due to the infection. These simulation results match those of numerical studies based on the mean field approximation (Figs. 8.2 and 8.3).

Table 8.5 Parameters for the simulations of the self-repair network with infection

Number of nodes	1000
Initial number of normal nodes	500
Number of steps	500
Repair rate P_r	0.00–1.00 (in 0.01 increments)
Repair success rate by normal nodes P_{rn}	1.0
Repair success rate by abnormal nodes P_{ra}	0.00–1.00 (in 0.01 increments)
Infection rate P_i	0.0, 0.5, 1.0

Fig. 8.4 Active phase (*left region* where some nodes remain abnormal) and frozen phase (*right region* where all the nodes are normal). The *solid* (*dotted*) *line* is the border separating two phases in mutual (self) repair

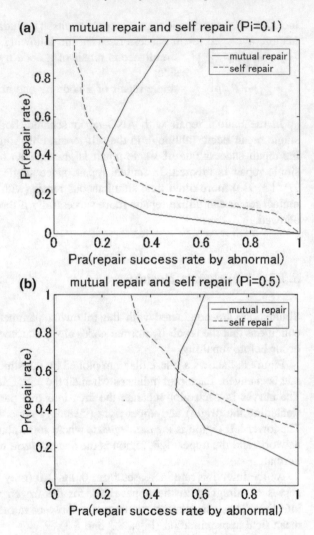

When self-repair and mutual (AND) repair are compared, it is observed that self-repair outperforms mutual repair when the repair rate is high, while mutual repair outperforms when the repair rate is low.

8.4 Discussion

From the simulation results, we observed that raising the infection rate P_i has the effect of dragging the border (separating active phase and frozen phase) toward the upper right. Thus, the impact of increasing P_i can be compensated by raising the repair rate P_r (or by raising P_{ra} if it were possible). Thus, infection is symmetric to repair in the sense it can be compensated. However, raising the infection rate affects the position of the border (affecting the ratio of P_i/P_{ra} at the border), while raising the repair rate cannot drag back the border. Thus, infection is intrinsically not symmetric to repair in the self-repair network.

It must be noted that the model with self-repair and infection still does not reflect real-world situations. In real computer networks, the network topology is neither necessarily a ring nor even a regular structure; several parameters such as infection rate, repair rate, and repair success rate are not uniform and so on; real-world situations are not that simple. However, a simple model has the merit of yielding some knowledge, which, although it cannot be directly applied to real situations, may be applied in many situations under several limiting conditions.

With these limitations in mind, the self-repair with infection reflects the current situation where an anti-virus program is installed in personal computers. Although anti-virus programs are often installed in server-client systems, which may be close to mutual (but one direction) repair, we do not consider the case here.

Since there are so many computer viruses both active and dormant, the infection rate is certainly not zero ($P_i > 0$).

Focusing on self-repair, the repair by a normal node will always succeed, hence the repair success rate by a normal node is 1 ($P_{rn} = 1$). Repair by an abnormal node may not succeed, and hence the repair success rate by an abnormal (infected) node is smaller than 1 ($P_{ra} < 1$). However, this parameter will decrease unless signature files are regularly updated (which imposes a cost on each computer). If the repair success rate by abnormal nodes (P_{ra}) is high enough, then the simulation indicates that mutual repair is preferable because it can eradicate abnormal nodes even with a lower repair rate (P_r) and hence low cost.

One difficulty of mutual repair is that distributed autonomy encourages nodes to ignore other nodes' problems (self-governance). However, a game theoretic approach with a systemic payoff that evaluates the costs and benefits throughout the system over the long term indicates that even for selfish agents, they will be more likely to cooperate (Chap. 4) (Ishida 2008).

8.5 Conclusion

Asymmetry requires two components each of which satisfies an asymmetric relation (e.g., a > b) to the other. This book adopts asymmetry as an important concept, as found in existence and non-existence property; normal and abnormal state; frozen and active phase. Two interactions between nodes, repair and infection, turned out to be asymmetric although the effects of the one can be compensated by the other in some cases.

As the self-repair network has the infection, it can deal with realistic situations In the following Chap. 9, we will see how the self-repair network with infection can be applied to the problem of information security.

References

Barabási, A.L., Freeh, V.W., Jeong, H.W., Brockman, J.B.: Parasitic computing. Nature **412** (6850), 894–897 (2001). doi:10.1038/35091039

Domany, E., Kinzel, W.: Equivalence of cellular automata to Ising models and directed percolation. Phys. Rev. Lett. **53**(4), 311–314 (1984)

Foster, I., Kesselman, C.: Computational grids—invited talk (Reprinted from The Grid: Blueprint for a new computing infrastructure, 1998). Lect. Notes Comput. Sci. **1981**, 3–37 (2001)

Foster, I., Kesselman, C.: The Grid 2: Blueprint for a New Computing Infrastructure. Morgan Kaufmann, (2003)

Foster, I., Kesselman, C., Tsudik, G., Tuecke, S.: A security architecture for computational grids. In: Proceedings of the 5th ACM Conference on Computer and Communications Security, pp. 83–92. ACM (1998)

Ishida, Y.: Complex systems paradigms for integrating intelligent systems: a game theoretic approach. In: Computational Intelligence: A Compendium, pp. 155–181. Springer, Berlin (2008)

Ishida, Y., Tanabe, K.-i.: Asymmetry in repairing and infection: the case of a self-repair network. In: Knowledge-Based and Intelligent Information and Engineering Systems, pp. 645–651. Springer, Berlin (2010)

Chapter 9
Dynamics of Self-Repair Networks of Several Types

Abstract We discuss the problem of spreading of the normal state (rather than spreading of the abnormal state), which is formalized as cleaning a contaminated network by mutual copying and self-copying. Repairing by copying is a double-edged sword problem that could spread contamination unless properly used. This chapter further focuses on transient states of self-repair networks involving several types of repair. Some implications for the framework of antivirus systems (against computer viruses and worms) will be presented comparing mutual repair and self-repair of nodes in the network.

Keywords Combination of repair types · Extended phase diagram · Transient phase diagram · Anti-virus program · Infection model · Probabilistic cellular automaton

9.1 Introduction

Intra-computer technology such as multi-core processors as well as inter-computer technology such as networking computers have made computer systems more complex and large-scale. The technological progress of computers not only allows but also requires computers to monitor themselves in an intra- (Brown and Patterson 2001; Brown 2004), (Hoffmann et al. 2010) and inter-computer fashion (Foster and Kesselman 2003). As human nervous systems created consciousness to monitor themselves (and multi-cellular organisms created immune systems to monitor themselves), computers, after exceeding some threshold of complexity, would need self-recognition and self-reaction.

We proposed a self-repair network for networked computers and sensors/actuators to repair *the self* to deal with the network cleaning problem of

---

Most results of this chapter are presented in Ishida and Tanabe (2014).

© Springer International Publishing Switzerland 2015

Y. Ishida, *Self-Repair Networks*, Intelligent Systems Reference Library 101,

DOI 10.1007/978-3-319-26447-9_9

how to enable a collection of computers to repair themselves and clean up abnormal states (or reset the states of all the nodes).

This chapter further extends the self-repair network by combining two typical repairs: *self-repair* and *mutual repair*. Two types of hybridization, *mixed repair* and *switching repair*, will be investigated to broaden the parameter region (called *frozen phase*) where all the abnormal nodes are eradicated. Simulations and analyses based on the mean field approximation are used to draw a phase diagram that separates the frozen phase from the parameter region (called *active phase*) where some abnormal nodes remain.

Section 9.2 revisits the self-repair network involving infections. Two types of primitive repair of self-repair and mutual repair as well as two types of their combination, mixed repair and switching repair, are presented. Section 9.3 examines the model with a steady-state analysis and a computer simulation for these repairs. Section 9.4 presents a transient-state analysis by introducing variations of the phase diagrams. Section 9.5 discusses the implications of the model and simulation targeting antivirus software under the restrictions assumed for the model. Constants calculated for steady-state analysis are collected together in the appendices.

The problem for this chapter is to observe transient states as well as stationary states of self-repair networks:

> How can we analyze and visualize the transient states as well as the stationary states of self-repair networks?

Self-repair networks with several types of networks and repair schemes are considered for the self-repair networks modeled by the probabilistic cellular automaton simplified from Chap. 3 (the same applies throughout Chaps. 7–10):

- Fault: A node becomes abnormal not only when the repair failed but also independently with a constant rate (failure rate).
- Repair: Not only normal nodes but also abnormal nodes can repair with a higher risk of unsuccessful repair. Repair does not require any cost.
- Control: The repair actions are executed on two neighbors simultaneously and independently with a constant rate (repair rate).
- Network: An N-node one-dimensional lattice with two neighbors and with the periodic boundary condition.

We have obtained results for the stationary states of self-repair networks by using a Markov chain with a mean field approximation and a pair approximation. This chapter extends the phase diagram to observe transient states.

9.2 Basic Model

9.2.1 Definitions and Extensions

The self-repair network consists of autonomous nodes capable of repairing neighbor nodes (i.e. connected nodes) by copying their contents. Each node has a binary state: normal (0) or abnormal (1). We further assume the simplest network: a ring structure in which the nodes are networked (Fig. 9.1).

In the self-repair network, repair may be divided into two types depending on the target of repair: *self-repair* (Fig. 9.1a), which targets the repairing node itself, and *mutual repair* (Fig. 9.1b), which targets the nodes connected to the repairing node. In the self-repair by each node, there is no interaction among nodes. As a result, only two cases take place: a normal node repairs a normal node, and an abnormal node repairs an abnormal node. The other two cases—a normal node repairs an abnormal node and an abnormal node repairs a normal node—would not take place in the self-repair.

Each node repairs the neighbor nodes with a probability P_r (called *repair rate*). The repair will succeed with a probability P_{rn} (called *repair success rate by normal nodes*) when it is done by a normal node, but with a probability P_{ra} (called *repair success rate by abnormal nodes*) when done by an abnormal node. Infection occurs only when at least one infected node exists in the neighborhood. Infection occurs with a probability P_i (called *infection rate*).

When many repairs are applied to one node simultaneously (called *simultaneous repair*), the repair can be further subdivided into two types: AND-repair and OR-repair. In the AND-repair, all the repairs must be successful when the repairs are done by multiple nodes simultaneously, while the OR-repair requires at least one repair to be successful out of multiple repairs done simultaneously.

We can control only the repair rate P_r but not the other parameters: the repair success rates P_{rn} and P_{ra}, and the infection rate P_i. These parameters are determined based on the environment of the network and computers. We cannot choose

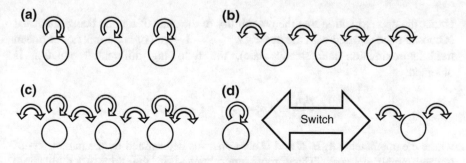

Fig. 9.1 **a** Self-repair; **b** Mutual repair; **c** Mixed repair; **d** Switching repair

between the AND-repair or OR-repair where they provide the theoretical worst case (AND-repair) and the best case (OR-repair). In reality, the situation will lie between the two extremes depending on the protocol and synchrony of the network.

9.2.2 Combining Self-Repair and Mutual Repair

The two types of repair, *self-repair* (Fig. 9.1a) and *mutual repair* (Fig. 9.1b), can be combined in two ways: *mixed repair* (Fig. 9.1c) and *switching repair* (Fig. 9.1d). The mixed repair just uses both self-repair and mutual repair, while the switching repair switches these two repairs. That is, each node executes the self-repair with a switching rate P_{sr}, and the mutual repair with $1 - P_{sr}$. For the switching repair, mutual OR-repair has a large area of frozen phase and need not be switched to the self-repair. Therefore, we consider only AND-repair for the switching repair.

Although mutual repair and switching repair can involve up to two simultaneous repairs (Fig. 9.1b, d) and mixed repair up to three (Fig. 9.1c), self-repair includes only the single repair (Fig. 9.1a). In AND (OR)-repair, because the repair success rate of single (simultaneous) repair is greater than that of simultaneous (single) repair, single (simultaneous) repair is favored. Since the higher the repair rate P_r, the higher the probability of simultaneous repair, a smaller (larger) repair rate is favored for AND (OR)-repair.

The transition probabilities for probabilistic rules are listed in the appendices: Table A.1 (mutual repair), Table A.3 (mixed repair) and Table A.5 (switching repair). Those for self-repair are listed in Table 8.3 in Chap. 8 (Ishida and Tanabe 2014, 2010).

9.3 Analyses and Simulations

9.3.1 Steady-State Analysis

Under the approximation that the probability of the state of a node being abnormal is constant and equated with the density ρ_1 ($\rho_0 = 1 - \rho_1$) of abnormal nodes (mean field approximation and steady state), the following differential equation is obtained:

$$\frac{d\rho_1}{dt} = A\rho_1^3 + B\rho_1^2 + C\rho_1 + D,$$

where the coefficients A, B, C and D are constants determined by the parameters of the self-repair network. Three roots are obtained in the third-order algebraic equation given by the steady state, that is, the roots of $A\rho_1^3 + B\rho_1^2 + C\rho_1 + D = 0$. These three roots are fixed points of the differential equation above, and we can obtain the density of abnormal nodes in the steady state determined by the stable point corresponding to the root. Figures 9.2 and 9.3 plot the density obtained by a numerical study of the roots of the above algebraic equation.

Coefficients A, B, C and D of the equation expressed by parameters for several types of repair discussed in this chapter are listed in Table A.2 (mutual repair), Table A.4 (mixed repair), and Table A.6 (switching repair) in the appendices. Those for the self-repair are listed in Table 8.4 in Chap. 8. Both mutual repair and mixed

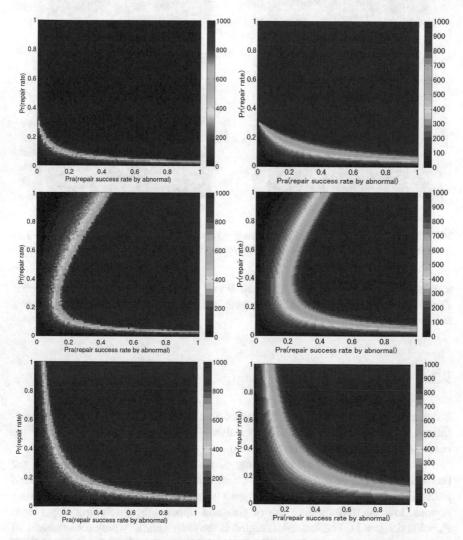

Fig. 9.2 Density of normal nodes in the steady state with $P_i = 0.1$. By simulation (*left*) and by mean field approximation (*right*); self-repair (*bottom row*), mutual AND-repair (*middle row*), mutual OR-repair (*top row*)

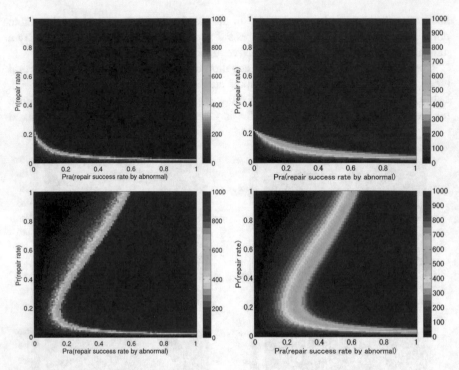

Fig. 9.3 Density of normal nodes in the steady state with $P_i = 0.1$. By simulation (*left*) and by mean field approximation (*right*); mixed AND-repair (*below*) and mixed OR-repair (*above*)

repair are further divided into AND-repair and OR-repair. For any types of repair, $D = 0$ when we limit ourselves to the case that the repair by normal nodes always succeeds ($P_{rn} = 1$).

9.3.2 Simulation Results

Simulations are conducted with the parameters listed in Table 9.1 under the conditions that the repair by normal nodes always succeeds ($P_{rn} = 1$) and infections occur ($P_i = 0.1$) before repairing ($P_r > 0$).

Figures 9.2 and 9.3 show the density of normal nodes, which is similar to the phase diagram but including detailed information on any region (called *extended phase diagram*). The curves in each plot separate the area into two parameter regions: lower-left (including the origin) and upper-right (including upper-right $P_r = 1$ and $P_{ra} = 1$). The lower-left region is the *active phase* where some abnormal nodes remain in the network, and the upper-right region is the *frozen phase* where all the nodes become normal.

Table 9.1 Parameters for the simulations of the self-repair network with infection

Number of nodes	1000
Initial number of normal nodes	500
Number of steps	500
Repair rate P_r	0.00–1.00 (in 0.01 increments)
Repair success rate by normal nodes P_m	1.0
Repair success rate by abnormal nodes P_{ra}	0.00–1.00 (in 0.01 increments)
Infection rate P_i	0.0, 0.1

The shape of the border between two phases may be explained qualitatively. Generally, there is a trade-off between the repair rate (P_r plotted on the vertical axis of the diagram) and the repair success rates (P_{ra} plotted on the horizontal axis of the diagram), for the decrease of the repair success rate must be compensated by the increase of the repair rate to attain the same level of eradication of abnormal nodes. As observed in Fig. 9.2, the trade-off is obvious for the self-repair which does not have repairs between nodes. When there are repairs between nodes, OR-repair gives the best bound, for only one success of the repair (among the repairs done by the two neighbor nodes) suffices for a node to be repaired, while AND-repair gives the worst bound, for all the repairs must be successful. For this reason, increasing the repair rate of AND-repair would lead to the contrary of the eradication of abnormal nodes (Fig. 9.2).

The extended phase diagram shows not only the active and frozen phases, but also the number of normal nodes in the steady state. The parameter regions are colored blue when normal nodes are dominant and red when abnormal nodes are dominant. Thus, the frozen phase is colored blue.

It can be observed that the frozen phase has one color because there are only normal nodes, while the color is graded in the active phase where the number of normal nodes decreases further away from the boundary.

The extended phase diagram obtained by simulations and by numerical studies based on the mean field approximation (Fig. 9.2) indicates that the density decreases as the parameter region goes further away from the upper-right where the eradication of abnormal nodes can be done most efficiently.

The density of normal nodes obtained by simulations qualitatively matches the density of normal nodes obtained by the mean field approximation (Figs. 9.2 and 9.3), but they do not match exactly (due to the approximation).

Regarding the comparison between the self-repair and the mutual AND-repair, Fig. 9.2 indicates that self-repair outperforms mutual AND-repair when the repair rate is high, while mutual AND-repair outperforms when the repair rate is low. Mutual OR-repair outperforms the self-repair when the repair rate is high, and its performance is similar to that of mutual AND-repair when the repair rate is low.

The higher performance (broader frozen region) of OR-repair compared with AND-repair when the repair rate is high can be observed both in the density of normal nodes (Figs. 9.2 and 9.3) and the time steps required to eradicate abnormal nodes (Fig. 9.7).

The shrinking of the frozen region caused by simultaneous repair is more severe in mixed AND-repair (Fig. 9.3) than mutual AND-repair (Fig. 9.2), for in a ring structure up to three simultaneous repairs can occur in the mixed AND-repair whereas up to two simultaneous repairs can occur in the mutual AND-repair.

Figure 9.4 shows the conventional phase diagrams for the switching repair with various parameters: $P_{sr} = 0.2, 0.5, 0.8$ (legend) and $P_i = 0.0$ (below), 0.1 (above). The effect of involving infection can be observed in the diagrams. The plots by mean field approximation (right) cannot be used to set an appropriate switching rate P_{sr} due to too much bias. The plots by simulations (left) suggest that there is a trade-off: if we increase P_{sr} (e.g. 0.8) then we get a broader frozen phase with the region P_r high, while we get a broader frozen phase with the region P_r low when we decrease P_{sr} (e.g. 0.2). This trade-off is obvious when the infection is involved (plots above).

Table 9.2 summarizes the characteristics of the four types of repair: self-repair, mutual repair, mixed repair and switching repair. Among the last three repairs having simultaneous repairs, we consider AND-repair and OR-repair for the mutual repair and the mixed repair, which makes six repairs in total. The mixed repair and switching repair will be considered in the next section.

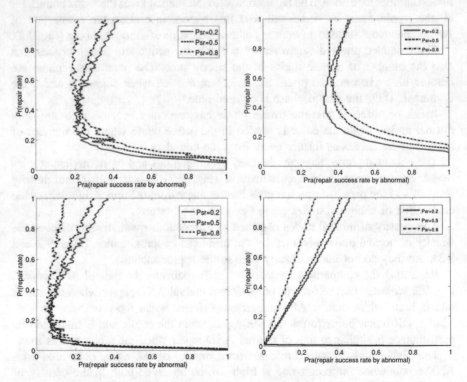

Fig. 9.4 Active phase (*left region* where some nodes remain abnormal) and frozen phase (*right region* where all the nodes are normal) in switching AND-repair. By simulation (*left*) and by mean field approximation (*right*); with $P_i = 0.0$ (*below*) and $P_i = 0.1$ (*above*)

Table 9.2 Summary of the characteristics of each repair

Repair type	Parameter characteristics	Simultaneous repair
Self	Higher performance with higher repair rate P_r	No simultaneous repair
OR (Mutual, Mixed)		Simultaneous repair
AND (Mutual, Mixed)	Higher performance with lower (but not zero) repair rate P_r	
AND Switching	The repair rate P_r can be adjusted by changing the switching rate P_{sr}	

9.4 Transient-State Analysis to Combine the Self- and Mutual Repair

To investigate the steady state, we can carry out not only simulations but also mathematical analysis based on spatial approximations such as mean field analysis and pair approximation (Sect. 10.2). To study the transient state, however, we can do only simulations; we use the parameters listed in Table 9.1 (infection rate is fixed: $P_i = 0.1$) for these simulations.

Figure 9.5 shows the extended phase diagrams of the four types of repair: self-repair, mutual repair, mixed repair and switching repair.

Figure 9.6 suggests that the number of normal nodes does not fluctuate in the parameter regions in the frozen phase such as ④. However, the number of normal nodes fluctuates in the active phase such as ① and ⑤, and fluctuates more as the parameters approach the boundary.

In practice, even if the parameters are in the frozen region and hence abnormal nodes can be eradicated, it would be meaningless if the eradication takes a very long time. Figure 9.7 shows the time steps required to eradicate the abnormal nodes to investigate the transient state before reaching the steady state. The diagram is again similar to the phase diagram but includes detailed information on the eradication time required (called *transient phase diagram*). The transient phase diagram shows not only the active and frozen phases, but also the number of steps required to reach the steady state. The parameter regions are colored blue when the number of steps is small and red when large. Thus, the frozen phase (*right region* where all the nodes are normal) is colored blue.

In contrast to an extended phase diagram which has one color in the frozen phase and graded in the active phase, the transient phase diagram has one color in the active phase (infinite number of steps required to eradicate abnormal nodes) while the color is graded in the frozen phase where the number of steps required decreases further away from the boundary.

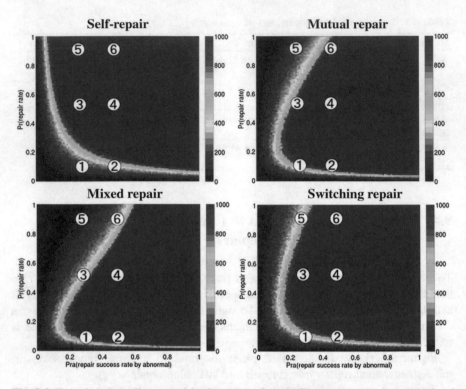

Fig. 9.5 Extended phase diagrams of the four types of repair. The parameter regions are colored *blue* when normal nodes are dominant and *red* when abnormal nodes are dominant

It can be observed from Fig. 9.7 that within the frozen phase, the number of steps required to eradicate the abnormal nodes differs. The *top-right* region requires the least steps, and the number of steps required for eradication increases as the parameter region goes further away from this region. Put another way, abnormal nodes are eradicated faster when the repair efficiency is high (both parameters P_r and P_{ra} are high in this region).

9.5   Discussion: Implications for Antivirus Software

It has been pointed out that the critical point (above which an epidemic breaks out) in scale-free networks is 0, thus it is difficult to eradicate computer viruses and worms from the Internet (Dezso and Barabási 2002), which may be a scale-free network (Barabási and Frangos 2002). However, from the viewpoint of the self-repair network discussed above, the critical point is not 0, and computer viruses and worms may be eradicated if the network involves an active repair which has a certain probability of repair success.

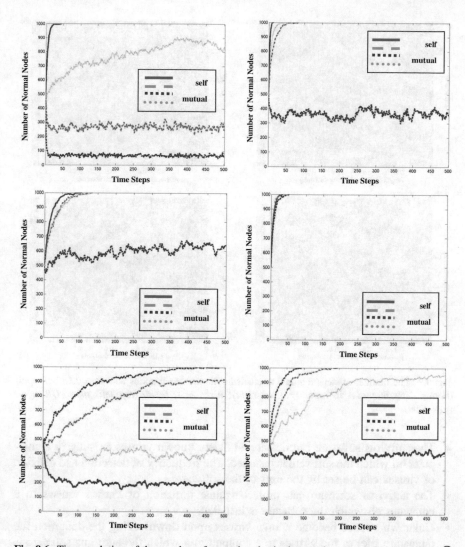

Fig. 9.6 Time evolution of the number of normal nodes in the transient state with parameters ① $(P_r = 0.1, P_{ra} = 0.3)$ (*bottom row, left*); ② $(P_r = 0.1, P_{ra} = 0.5)$ (*bottom row, right*); ③ $(P_r = 0.5, P_{ra} = 0.3)$ (*middle row, left*); ④ $(P_r = 0.5, P_{ra} = 0.5)$ (*middle row, left*); ⑤ $(P_r = 0.9, P_{ra} = 0.3)$ (*top row, left*); ⑥ $(P_r = 0.9, P_{ra} = 0.5)$ (*top right, left*)

As an application of the self-repair network, we examine the current antivirus software (against computer viruses and worms, file contamination viruses in particular) and consider the potential of several types of repair. It should be stressed that the self-repair network is a model to access the theoretical worst (or best) bounds, and is not a model that will directly lead to new antivirus software. Here, antivirus software is defined as follows:

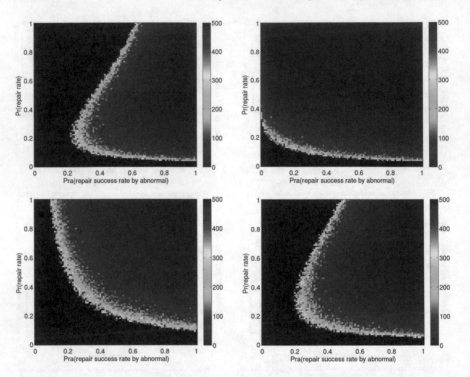

Fig. 9.7 The number of steps required to eradicate the abnormal nodes with $P_i = 0.1$. *Self repair* (*bottom, left*); *mutual AND repair* (*bottom, right*); *mixed AND repair* (*top, left*); *mixed OR repair* (*top, right*)

- The antivirus software can detect and delete known viruses existing in a computer on which the software is installed. The frequency of detection and deletion of viruses can be set by the user of the software.
- The antivirus software can protect against intrusion of known viruses to a computer on which the software is installed.
- Unknown viruses become known viruses upon downloading the definition file (signature file) of the viruses to a computer on which the antivirus software is installed.

Further, computers on which the software is installed are networked in a ring form, which is the topology assumed for the self-repair network discussed in this chapter.

If we view antivirus software from the self-repair network, the binary state of normal or abnormal of the self-repair network respectively correspond to the state without viruses and the state with viruses in computers on which the software is installed. Repair in the self-repair network corresponds to the detection and deletion of viruses in computers.

Among the parameters in the self-repair network, P_r (repair frequency rate), P_{rn} (repair success rate by normal nodes), P_{ra} (repair success rate by abnormal nodes), and P_i (infection rate), only P_r can be set by the user of the antivirus software. This P_r corresponds to the frequency of virus detection/deletion in the software. Frequent detection/deletion by the antivirus software on a computer requires computational resources (CPU time and memory) and could interrupt or hamper the task of the computer. The other parameters P_{rn}, P_{ra} and P_i depend on the performance of not only the antivirus software but also that of the computer environment including the computer, computer network and viruses, and hence cannot be set by the antivirus software.

The repair success rate by normal nodes P_{rn} corresponds to a success rate of detection/deletion of known viruses (i.e. the virus definition file is downloaded onto the computer) by a *normal* computer (a computer not infected by viruses). Usually, an uninfected computer on which the antivirus software and definition file are installed can successfully detect and delete viruses, hence P_{rn} can be assumed to be one ($P_{rn} = 1$).

The repair success rate by abnormal nodes P_{ra} corresponds to the success rate of detection/deletion of known viruses by an *abnormal* computer (a computer infected by the viruses). Even an infected computer (on which the antivirus software and definition file are installed) may be able to detect and delete a virus successfully if the virus definition file is properly installed and the function of detection/deletion is not hampered by the virus, thus P_{ra} can be assumed to be greater than zero ($0 < P_{ra} < 1$).

The infection rate P_i is the rate at which the virus successfully enters a computer from outside (through the network or other removable media). This rate depends upon whether the virus definition file has already been downloaded or not. The infection rate is very low if the definition file exists in the computer, but high otherwise.

Since the current antivirus software detects/deletes viruses in the computer in which it is installed, the type of repair is *self-repair* rather than *mutual repair*. Thus, the corresponding phase diagram is a diagram of self-repair with infection rate greater than zero (Fig. 9.5). For a known virus, the infection rate is small (but not zero) and the repair success rate by normal nodes (P_{rn}) and that by abnormal nodes (P_{ra}) are high. The phase diagram in Fig. 9.5 (e.g., place indicated by the circled number 2) suggests that the viruses can be eradicated even with low (but not zero) frequency of repair (P_r). For unknown viruses, however, a lower repair success rate by abnormal nodes (P_{ra}) would require a higher frequency of repair (P_r) for eradication (e.g., place indicated by the circled number 3 instead of 1 in Fig. 9.5).

Existing antivirus systems adopt *self-repair*, which requires a high frequency of repair (P_r), however, a higher frequency of repair is not favorable due to the larger computational resources required. To solve this problem, let us consider other types of repair in the context of existing antivirus systems.

For the parameter region with a high frequency of repair (P_r), *self-repair* would be favorable (as observed in Fig. 9.5 comparing the places indicated by the circled numbers 3 and 5 to those of other types). Beside the performance, *self-repair* does not impose a psychological burden, for it is an independent system and would not bother other computers in repairing and being repaired.

For a parameter region with low frequency of repair (P_r), the *mutual repair* and *mixed repair* would be favorable (as observed in Fig. 9.5 comparing the places indicated by the circled numbers 1 and 2 to those of *self repair*). When the repair is *networked* as in *mutual repair* and *mixed repair* (Fig. 9.1), simultaneous repair by multiple computers occurs. Then distinction between AND-repair and OR-repair should be taken into consideration. Again, it depends on the environment of the computer network and computers and is not under our control (it could be controlled if we were to design the protocol for repairing). Generally, OR-repair may be favorable, because its broader area of the frozen phase and the mutual repair need not be switched to the self-repair if OR-repair is available (by avoiding simultaneous repair).

Since it would be impossible to evaluate the parameters (including AND-repair, OR-repair) beforehand and the parameters change dynamically, another option would be the *switching repair*. The switching parameter P_{sr} can be set depending on the other parameters so that each frozen region of the repairs involved together will cover a broader and practically important area of the phase diagram.

The above discussion does not consider the problem of privacy. If we place importance on privacy, the *self-repair* would be favorable compared with repairs involving *mutual repair* such as *mutual, mixed* and *switching repair*. If we restrict mutual repair to intranets within the home, companies and schools, or to cloud computing where clusters of computers and data centers are organized by one provider, then privacy would be less of a burden. More practically, the mutual repair may be restricted between servers of the antivirus provider and client computers, and the repairs may also be restricted to one way, from the servers to the clients. In fact, a *chain repair*, which may be placed as a dual of *chain failure,* is defined as a chain in which node A repairs node B and then node B in turn repairs node C and so forth. Evidently, in the case of chain repair starting from a normal node and assuming that P_{rn} equals one, then all the abnormal nodes will be eradicated.

For actual deployment in an existing network such as the Internet, more comprehensive simulations and analyses are required using a more realistic (dynamic larger scale and complex topology) network to test and compare these different types of repair.

If the topology is different from a ring structure, the *degree* of each node changes and this change of *degree* makes the possible number of simultaneous repairs change. In fact, it may be possible to avoid simultaneous repairs by designing a protocol for when multiple repairs arrive at the same node and at the same time. The protocol could randomly choose one of the repairs and discard other repairs, or arrange their time of application in a serial fashion. In this case, success or failure of the repair would depend on the success or failure of the chosen one (for the randomly chosen repair) and the last applied one (for the serially arranged repair).

9.6 Conclusion

Although other chapters concentrated on the stationary state, this chapter focused on the dynamics of the stationary state, that is, the transient state. The dynamics of the self-repair network are of vital importance, particularly when dealing with engineering problems, for it is meaningless if it takes a very long time to reach the stationary state even if it exists. We have visualized the transient state by devising the phase space diagram in two ways: extended phase diagram (showing the density of normal nodes in the active phase), and transient phase diagram (showing the time steps required to eradicate abnormal nodes in the frozen phase). It is demonstrated that these two types of variant phase diagram play an important role in combining several types of self-repair and tuning the parameters of self-repair networks.

In the following Chap. 10, we will see how the self-repair network can be related to existing epidemic models.

References

Barabási, A.-L., Frangos, J.: Linked: The New Science of Networks Science of Networks. Basic Books (2002)

Brown, A.: Recovery-oriented computing: building multitier dependability (2004)

Brown, A., Patterson, D.A.: Embracing failure: a case for recovery-oriented computing (roc). In: High Performance Transaction Processing Symposium, pp. 3–8 (2001)

Dezso, Z., Barabási, A.L.: Halting viruses in scale-free networks. Phys. Rev. E. **65**(5) (2002). doi: Artn 055103; 10.1103/Physreve.65.055103

Foster, I., Kesselman, C.: The Grid 2: Blueprint for a New Computing Infrastructure. Morgan Kaufmann (2003)

Hoffmann, H., Maggio, M., Santambrogio, M.D., Leva, A., Agarwal, A.: Seec: a framework for self-aware computing (2010)

Ishida, Y., Tanabe, K.-i.: Asymmetry in repairing and infection: the case of a self-repair network. In: Knowledge-Based and Intelligent Information and Engineering Systems, pp. 645–651. Springer, Berlin (2010)

Ishida, Y., Tanabe, K.-i.: Dynamics of Self-Repairing Networks: Transient State Analysis on Several Repair Types. International Journal of Innovative Computing, Information and Control **10**(1) (2014)

Chapter 10
Self-Repair Networks as an Epidemic Model

Abstract This chapter focuses on propagation and spreading state caused by self-repairing or by infection. We proposed a self-repair network where nodes are capable of repairing neighbor nodes by mutual copying, and investigated a critical point at which faulty nodes can be eliminated. This chapter further studies the dynamics of eradicating abnormal nodes by comparing the self-repair network with mathematical epidemic models such as SIS models. It is shown that the self-repair network, which is a probabilistic cellular automaton, can be regarded as an epidemic model in some restricted situations. Since the self-repair network is related to the epidemic models by corresponding parameters, this chapter also serves to explain the calculus of how the network cleaning threshold is derived by the mean field approximation.

Keywords Epidemic model · SIS model · SIR model · Vaccination · Infection model · Probabilistic cellular automaton

10.1 Introduction

Information networks are complex systems with multiple and interacting processes acting in parallel on various network structures. Self-recovery of such networks has been studied with a focus on the network structure (Dezso and Barabási 2002) as well as interacting processes (Brown and Patterson 2001). We have been studying a self-repair network focusing on state propagation and regulation, while keeping the network model simple enough for analysis, as described in Chap. 3 (Ishida 2005, 2008). The model involves two state propagations: the abnormal state propagation by unsuccessful repair and the normal state propagation by successful repair.

Some results of this chapter are presented in Aoki and Ishida (2008).

© Springer International Publishing Switzerland 2015

Y. Ishida, *Self-Repair Networks*, Intelligent Systems Reference Library 101,
DOI 10.1007/978-3-319-26447-9_10

Epidemic models have been studied for a long time, and their nonlinear properties have been investigated in great detail. The models include not only those described by differential equations (Lajmanovich and Yorke 1976) but also by probabilistic cellular automata (Rhodes and Anderson 1996) or even including moving agents (Boccara and Cheong 1993). On the other hand, phase transitions have been studied on models extended from the Ising model in the field of statistical physics but involving probabilistic cellular automata, i.e. the Domany-Kinzel model (1984).

Our model has already been related to a model in statistical physics (Ishida 2005). This chapter specifically focuses on the relation between the self-repair network and an epidemic model called the SIS model. The self-repair network is related to the epidemic models by corresponding parameters, as was done for the Domany-Kinzel model in Chap. 3.

The problem for this chapter is to relate self-repair networks to the epidemic model:

Can self-repair networks be related to the SIS model?

The self-repair network to be related here is the simplest one modeled by the probabilistic cellular automaton simplified from Chap. 3 (the same applies throughout Chaps. 7–10):

- Fault: A node becomes abnormal not only when the repair failed but also independently with a constant rate (failure rate).
- Repair: Not only normal nodes but also abnormal nodes can repair with a higher risk of unsuccessful repair. Repair does not require any cost.
- Control: The repair actions are executed on two neighbors simultaneously and independently with a constant rate (repair rate).
- Network: An N-node one-dimensional lattice with two neighbors and with the periodic boundary condition.

10.2   Parameters in the Self-Repair Network

10.2.1   A Summary of the Self-Repair Network

For comparison with epidemic models, the features of the simplest model of self-repair networks can be summarized as:

- Active Agents: The active aspect of agents capable of repairing is involved.
- Dual Nature: Agents are not only capable of repairing but also being repaired.

- Existence Asymmetry: Agents have an asymmetric nature of abnormal and normal, whose asymmetric nature comes from the asymmetric nature of the existence property (or equivalently, the special feature of zero, i.e. non-existence).

The self-repair network is a network consisting of nodes with binary state of normal (0) or abnormal (1), each with the ability to repair neighbor nodes by copying its own content. Each node repairs the neighbor nodes with a probability P_r. The repair will be successful with a probability P_{rn} when it is done by a normal node, but with a probability P_{ra} when done by an abnormal node. Further, all the repairs must be successful for the target node to be normal when repairing is done by multiple nodes as in Fig. 10.1 (AND-repair).

When repairing is done by copying, the key difference from conventional repairing is that the repairing could contaminate rather than clean other nodes. This is the double-edged sword problem in repairing. Thus, it is important to investigate the conditions under which the network can be cleaned. To clarify the conditions, we used a probabilistic cellular automaton in Chap. 3 (Ishida 2005, 2008) which turned out to be the Domany-Kinzel model (1984) in some particular cases.

In a one-dimensional structure with two adjacent neighbor nodes (Fig. 10.2), the probabilities for each rule of state change are listed again in Table 10.1 (from Table 7.1 with coefficients to be used).

Fig. 10.1 Repair success rate when repairing is done by normal node (*above, left*) and by abnormal node (*above, right*); all the repairs by neighbor nodes must be successful for the target node to be normal (*below*). *Solid arcs* indicate repairing and *dotted arcs* indicate state change

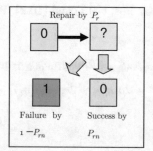
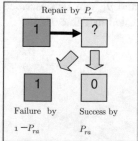

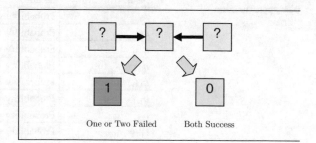

Fig. 10.2 One-dimensional structure with two adjacent neighbor nodes

Table 10.1 Transition probabilities of the one-dimensional probabilistic cellular automaton

State change	Transition probability	
$000 \rightarrow 1$	$P_r(1 - P_{rm})(P_r P_{rm} - P_r + 2)$	a
$001 \rightarrow 1$	$P_r^2(1 - P_{rm}P_{ra}) + P_r(1 - P_r)((1 - P_{rm}) + (1 - P_{ra}))$	b
$101 \rightarrow 1$	$P_r(1 - P_{ra})(2 - P_r(1 - P_{ra}))$	c
$010 \rightarrow 0$	$P_r P_{rm}(2(1 - P_r) + P_r P_{rm})$	d
$011 \rightarrow 0$	$P_r((P_{rm} + P_{ra})(1 - P_r) + P_r P_{ra} P_{rm})$	e
$111 \rightarrow 0$	$P_r P_{ra}(P_r P_{ra} + 2(1 - P_r))$	f

10.2.2   Steady-State Analysis with Pair Approximation

Pair approximation (e.g., Dieckmann et al. 2000, and references therein) can calculate dynamics by using the probabilities listed in Table 10.2. The dynamics of ρ_1 and ρ_{11} can be described by Eqs. (10.1) and (10.2). Parameters a, b, c, d, e, f are listed in Table 10.1.

$$
\begin{aligned}
\frac{d\rho_1}{dt} = {} & a\rho_0 q_{0/0}^2 + 2b\rho_0 q_{1/0}q_{0/0} + c\rho_0 q_{1/0}^2 \\
& - d\rho_1 q_{0/1}^2 - 2e\rho_1 q_{0/1}q_{1/1} - f\rho_1 q_{1/1}^2
\end{aligned}
\tag{10.1}
$$

Table 10.2 Probability of pair approximation

Name	Mean
ρ_0	Probability of normal nodes (0)
ρ_1	Probability of abnormal nodes (1)
$q_{0/0}$	Probability with (0) next to (0)
$q_{0/1}$	Probability with (0) next to (1)
$q_{1/0}$	Probability with (1) next to (0)
$q_{1/1}$	Probability with (1) next to (1)
ρ_{00}	Probability of pair (0)–(0)
$\rho_{01}(=\rho_{10})$	Probability of pair (0)–(1)
ρ_{11}	Probability of pair (1)–(1)

$$\frac{d\rho_{11}}{dt} = a^2\rho_{00}q_{0/0}^2 + 2ab\rho_{00}q_{0/0}q_{1/0} + 2b(1-d)\rho_{01}q_{0/0}q_{0/1}$$
$$+ 2b(1-e)\rho_{01}q_{0/0}q_{1/1} + 2c(1-d)\rho_{10}q_{0/1}q_{1/0} + 2b^2\rho_{00}q_{1/0}^2 \qquad (10.2)$$
$$+ 2c(1-e)\rho_{01}q_{1/0}q_{1/1} - 2(e+f-ef)\rho_{11}q_{0/1}q_{1/1}$$
$$- e(2-e)\rho_{11}q_{0/1}^2 - f(2-f)\rho_{11}q_{1/1}^2$$

when $\frac{d\rho_1}{dt} = 0$ and $\frac{d\rho_{11}}{dt} = 0$, the steady state can be calculated by Newton's method.

10.2.3 Steady-State Analysis with Mean Field Approximation

The above pair approximation is a detailed version of the mean field approximation (MFA), and hence can be reduced to the MFA when the pair-wise correlation is neglected: $q_{0/0} = \rho_0$, $q_{0/1} = \rho_0$, $q_{1/0} = \rho_1$, $q_{1/1} = \rho_1$. Then the dynamics of ρ_1 can be described by Eq. (10.3), which matches the equation obtained by the MFA in Sect. 7.3.1.

$$\frac{d\rho_1}{dt} = a\rho_0^3 + 2b\rho_0^2\rho_1 + c\rho_0\rho_1^2$$
$$- d\rho_0^2\rho_1 - 2e\rho_0\rho_1^2 - f\rho_1^3 \qquad (10.3)$$

The following Eq. (10.4) can be obtained by arranging Eq. (10.3) where A, B and C are constants determined by the three parameters of the self-repair network.

$$\frac{d\rho_1}{dt} = A\rho_1^3 + B\rho_1^2 + C\rho_1 + D \qquad (10.4)$$

$$B = -P_r^2(P_{rm} - P_{ra})^2 \qquad (10.5)$$

$$C = -2P_r(1 - P_{rm})(P_r(P_{rm} - P_{ra}) + 1) + P_r(P_r - 2P_{ra}) \qquad (10.6)$$

$$D = P_r(1 - P_{rm})(2 - P_r(1 - P_{rm})) \qquad (10.7)$$

The coefficient A can be canceled out to be 0. In order for abnormal nodes to be eradicated, D must be 0, i.e. $P_{rm} = 1$, for otherwise normal nodes could spread abnormal states. When $D = 0$, the following condition (10.8) must be satisfied for abnormal nodes to be eradicated (matching the result of Sect. 3.5, that is, the repair

success rate by abnormal nodes must exceed half of the repair rate), since the time
derivative $\frac{d\rho_1}{dt}$ must be negative in Eq. (10.4).

$$\frac{P_{ra}}{P_r} \geq \frac{1}{2} \tag{10.8}$$

10.2.4   Simulation Results

The dynamics of a normal node population must be investigated and verified by
computer simulation. Simulations are conducted with the parameters listed in
Table 10.3. Figure 10.3 plots the number of normal nodes for varying repair success

Table 10.3 Simulation
parameters

Parameter	Value
Time steps for each trial	1500
Number of trials	10
Number of nodes	400
Initial number of normal nodes	200
P_r	0.3, 0.6, 1.0
P_{rn}	1.0
P_{ra}	0–1 (0.02)

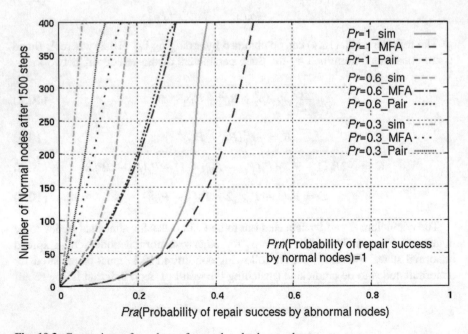

Fig. 10.3 Comparison of numbers of normal nodes by steady state

rate P_{ra} when repaired by abnormal nodes. It can be observed that the results by the pair approximation are closer to those by the (agent-based) simulations than those by the mean field approximation when the parameter P_r becomes smaller (closer to 0).

10.3   Self-Repair Network and SIS Model

10.3.1   An SIS Model

Epidemic models assume states such as S (susceptible), I (infected), and R (recovered). Combining these states, types of epidemic model include SI, SIS, and SIR. Among them, the SIS model assumes that susceptible nodes will fall into the infected state with an infection rate β when the neighbor nodes are infected. The infected nodes will be recovered with a recovery rate γ and enter the susceptible state again. The SIS model can describe, for example, sexually transmitted diseases such as gonorrhea (Lajmanovich and Yorke 1976).

Susceptible (S) and infected (I) state respectively correspond to normal (0) and abnormal (1) state in the self-repair network. Figure 10.4 shows the state transition between S (0) and I (1).

10.3.2   Parameter Correspondence Between an SIS Model
and a Self-Repair Network

In a random graph with a mean degree $\langle k \rangle$, the dynamics of the fraction of infected (abnormal) nodes ρ_1 can be described as follows.

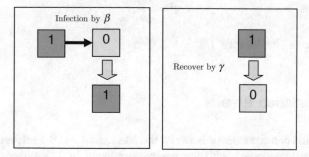

Fig. 10.4 State transition in an SIS model. *Solid arcs* indicate repairing and *dotted arcs* indicate state change

$$\frac{d\rho_1}{dt} = \beta\langle k\rangle(1 - \rho_1)\rho_1 - \gamma\rho_1$$
$$= -\beta\langle k\rangle\rho_1^2 + (\beta\langle k\rangle - \gamma)\rho_1 \tag{10.9}$$

This master equation has two equilibria: $\rho_1 = 0, 1 - \gamma/(\langle k\rangle\beta)$. Only $\rho_1 = 1 - \gamma/(\langle k\rangle\beta)$ is a stable equilibrium; $\rho_1 = 0$ cannot be attained unless it starts from $\rho_1 = 0$, otherwise the density ρ_1 converges on $\rho_1 = 1 - \gamma/(\langle k\rangle\beta)$ (starting from $\rho_1 > 0$). When $\gamma/\beta < 1/\langle k\rangle$, which will make $\frac{d\rho_1}{dt} < 0$ and $\rho_1 < 0$, the infected nodes are eradicated.

Thus, when $P_{rn} = 1$ and hence $D = 0$ in Eq. (10.4), parameters B and C of a self-repair network will be related to the parameters of the above SIS model as follows.

$$B = -\beta\langle k\rangle \tag{10.10}$$

$$C = \beta\langle k\rangle - \gamma \tag{10.11}$$

Since we consider a one-dimensional cellular automaton for the self-repair network, $\langle k\rangle$ can be considered as 2 (as in Fig. 10.2), and hence

$$\beta = \frac{P_r^2(1 - P_{ra})^2}{2} \tag{10.12}$$

$$\gamma = P_r P_{ra}(P_r P_{ra} + 2(1 - P_r)) \tag{10.13}$$

For the SI model, nodes once infected (I) cannot become susceptible (S) (see Fig. 10.4 to compare it with SIS), so the SI model can be considered as a special case of the SIS model where the recovery rate γ is identical to 0. Thus, when $P_{ra} = 0$ the self-repair network can be reduced to the SI model with infection rate:

$$\beta = \frac{P_r^2}{2},$$

which is always smaller than 1/2.

10.4   Simulation Results

To examine the correspondence between the SIS model and the self-repair network, simulations are conducted with the parameters listed in Table 10.4. Figure 10.5 plots the number of normal nodes for varying repair success rate P_{ra} when repaired by abnormal nodes. The SIS model and the self-repair network mostly match, but do not match the numerical solution obtained from the mean field approximation (1), particularly when P_{ra} is > 0.3.

Table 10.4 Simulation parameters

Parameter	Value
Time steps for each trial	1500
Number of trials	10
Number of nodes	400
Initial number of normal nodes	200
P_r	1.0
P_{rn}	1.0
P_{ra}	0–1 (0.02)
β	$\frac{P_r^2(1-P_{ra})^2}{2}$
γ	$P_r P_{ra}(P_r P_{ra} + 2(1 - P_r))$

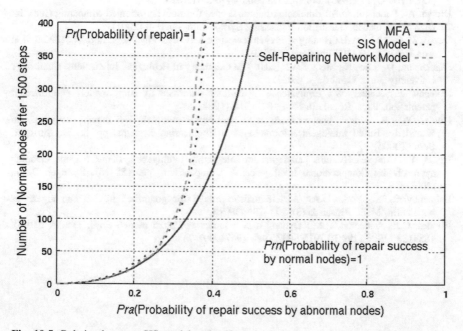

Fig. 10.5 Relation between SIS model and self-repair network

10.5 Conclusion

The self-repair network involves repairing by mutual copying in a network. Hence, it models not only abnormal state propagation (unsuccessful repair) but also normal state propagation (successful repair). Thus, it differs from epidemic models, which model only abnormal state propagation (infection), and recovery is modeled as an independent event (not as propagation). This chapter showed that, in spite of the difference, the self-repair network can be reduced to an SIS model when the repair by normal nodes always succeeds ($P_{rn} = 1$), and can be mapped to an SI model (provided the infection rate does not exceed 1/2) when repair by abnormal nodes always fails ($P_{ra} = 0$).

References

Aoki, Y., Ishida, Y.: Epidemic models and a self-repairing network with a simple lattice. Artif. Life Robot. **12**(1–2), 153–156 (2008)

Boccara, N., Cheong, K.: Critical-behavior of a probabilistic-automata network sis model for the spread of an infectious-disease in a population of moving individuals. J. Phys. Math. Gen. **26** (15), 3707–3717 (1993). doi:10.1088/0305-4470/26/15/020

Brown, A., Patterson, D.A.: Embracing failure: a case for recovery-oriented computing (roc). In: High Performance Transaction Processing Symposium, pp. 3–8 (2001)

Dezso, Z., Barabási, A.L.: Halting viruses in scale-free networks. Phys. Rev. E. **65**(5), (2002). doi: Artn 055103; 10.1103/Physreve.65.055103

Dieckmann, U., Law, R., Metz, J.A.J. (eds): The Geometry of Ecological Interactions. Cambridge University press, Cambridge (2000)

Domany, E., Kinzel, W.: Equivalence of cellular automata to Ising models and directed percolation. Phys. Rev. Lett. **53**(4), 311–314 (1984)

Ishida, Y.: A critical phenomenon in a self-repair network by mutual copying. In: Knowledge-Based Intelligent Information and Engineering Systems, pp. 86–92. Springer, Berlin (2005)

Ishida, Y.: Complex Systems paradigms for integrating intelligent systems: a game theoretic approach. In: Computational Intelligence: A Compendium, pp. 155–181. Springer, Berlin (2008)

Lajmanovich, A., Yorke, J.A.: A deterministic model for gonorrhea in a nonhomogeneous population. Math. Biosci. **28**(3), 221–236 (1976)

Rhodes, C.J., Anderson, R.M.: Dynamics in a lattice epidemic model. Phys. Lett. A **210**(3), 183–188 (1996). doi:10.1016/S0375-9601(96)80007-7

Chapter 11
Self-Repair Networks
and the Self-Recognition Model

Abstract We have studied a self-repair network by mutual copying. This chapter further considers a model that involves mutual recognition among nodes before repairing. Although the repairing with recognition outperforms the repairing without recognition, in some cases recognition hampers cleaning. It is shown that there is an optimum frequency of mutual recognition for the best cleaning of the network. The interplay between recognition error and repair error is also discussed.

Keywords Self-recognition model · Recognition error · Repair error · Hybrid system · Recognition before repair

11.1 Introduction

Although the structure and topology of many networks, both artificial and natural, have been studied, as well as their structural properties such as "scale-free" or "small-world", few studies have examined the interaction in the network or "semantics" of the network (as opposed to syntax). For example, the "scale-free" network (Barabási and Frangos 2002) focuses on the spatial scale-freeness; however, temporal scale-freeness also seems to be important.

We have been studying the interaction of the network and proposed the framework of immunity-based systems, that is, the "semantics" of the network. In our framework, the network is double-sided and would raise the self-referential paradox in a flat logic without distribution, and hence subtle tuning is needed as in the immune system.

The drastic development of the characteristics of faults, as seen in malicious faults such as computer viruses, calls for the design paradigm of the immune system, in which the "self" adapts itself to the environment.

Some results of this chapter are presented in Ishida and Sugawara (2006).

© Springer International Publishing Switzerland 2015   133
Y. Ishida, *Self-Repair Networks*, Intelligent Systems Reference Library 101,
DOI 10.1007/978-3-319-26447-9_11

The self-nonself discrimination problem dealt with in the immune system would raise the self-referential problem as often mentioned in statements such as: "I am a liar" and "This statement is false." One way to resolve this paradox may be hierarchical logic or distribution of subjects; we use the latter approach.

Dividing the subjects and placing them inside the system has been often discussed in Autopoietic Systems (Maturana and Varela 1980) and other complex systems. Placing distributed subjects in the system implies that the subjects have only local and unlabeled information in solving problems. Also, and more importantly, the subject can be the object on which the subject operates and interacts.

It is still not clear what we can learn from the self and nonself discrimination in the immune system (e.g., Langman and Cohn 2000; Flower and Timmis 2007; Ishida 2004), however, elimination is done in practice and hence the double-sided property that the subject must be the object is imperative. Thus, the immune system is a double-edged sword.

The problem for this chapter is when can self-repair networks be combined properly with the self-recognition model:

When and how can self-repair networks be used with self-recognition model?

The self-repair network to be combined is the simplest one modeled by the probabilistic cellular automaton simplified from Chap. 3 (the same applies throughout Chaps. 7–10):

- Fault: A node becomes abnormal not only when the repair failed but also independently with a constant rate (failure rate).
- Repair: Not only normal nodes but also abnormal nodes can repair with a higher risk of unsuccessful repair. Repair does not require any cost.
- Control: The repair actions are executed on two neighbors simultaneously and independently with a constant rate (repair rate).
- Network: An N-node one-dimensional lattice with two neighbors and with the periodic boundary condition.

11.2   Self-Recognition Model and Dynamic Logics

This section briefly illustrates the self-recognition model as a self-action network where the action in this model is recognition. Self-recognition leads to the logical paradox called the "liar's paradox" or "Cretan liar paradox." The self-recognition model tries to relax the paradox by introducing mutual recognition in an autonomous distributed system (Fig. 11.1).

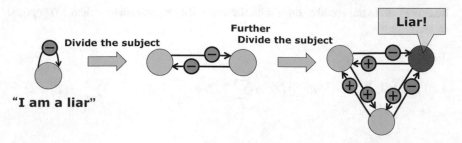

Fig. 11.1 Relaxing the liar's paradox by introducing mutual recognition in an autonomous distributed system. Nodes correspond to agents, and arcs with the + (−) sign mean the source agent says the target agent is honest (a liar)

Since the property of whether the node is abnormal or not is an asymmetric one related to the existence of the container rather than the state being contained, we need a higher logic than the logic of node states. For the purpose of computing and sensor network design, we introduce the concept of credibility in each node, which is expressed by a continuous dynamic variable normalized as a real value between 0 and 1. If the node state is fully reliable, the credibility should be 1, whereas if fully unreliable then the credibility must be set to 0. Asymmetry in the action is: fully unreliable agents may or may not lie (Byzantine fault).

$$\mathrm{d}x_i(t)/\mathrm{d}t = F\big(\{x_i(t)\}, \big\{a_{ij}\big(s_i(t), s_j(t), \mathit{aff}_{ij}(t)\big)\big\}\big),$$

- x_i: number of recognizing (or recognized) sets (T-cells, B-cells and antibodies)
- a_{ij}: interactions from type i to type j (positive for stimulation and negative for suppression)
- s_i: state of type i entity (e.g., activated/inactivated, virgin/immune, and so on)
- aff_{ij}: affinity between these two types.

Although credibility looks like the mathematical concept of probability, the only shared aspect is that the value is normalized to 0 to 1. The credibility does not have mathematical rigor such as the Bayesian network. For example, in mathematical probability (measure) all probabilities of all the exclusive events covering the entire events must add up to 1, while credibility does not have even the concept of exclusive events. For computation of credibility, the only important point is consistency among agent states. One way of computing credibility (or adjusting to the agent states) is to seek the credibility values that achieve the minimum inconsistency. This can be done by setting the total inconsistency analogous to energy (or a Lyapunov function) so that the total inconsistency decreases at each time step similarly to the Hopfield Network. This computation leads to a black and white model of the self-recognition model. There are several other engineering needs, e.g., ambiguous credibility may be reflected to credibility values (gray model), and credibility of ambiguous ones should be expressed as not credible ones with the credibility values smaller than 1 (skeptical model).

- Black & White : $\quad dr_i(t)/dt = \sum_j T_{ji}R_j + \sum_j T_{ij}R_j - \frac{1}{2} \sum_{j \in \{k:T_{ik} \neq 0\}} (T_{ij} + 1)$

- Gray : $\quad dr_i(t)/dt = \sum_j T_{ji}^+ R_j(t) - r_i(t)$

- Skeptical : $\quad dr_i(t)/dt = \sum_j T_{ji}^+ R_j(t)$

R_i credibility, which is the normalized value of r_i
r_i credibility before normalization
T_{ij} +1 (−1) for the arc from node i to node j with + (−) sign; 0 otherwise (for no arc)

11.3   Self-Repair by Copying: The Double-Edged Sword

11.3.1   A Model of the Self-Repair Network

This chapter concentrates on the naive problem of cleaning up the network by mutual copying: when can a contaminated network be cleaned up by mutual copying?

We have studied immunity-based systems and pointed out the possibility of the double-edged sword (Ishida 2004). For example, a system based on Jerne's idio-typic network framework has recognizing nodes that are also being recognized. Self-nonself recognition involving the self-referential paradox could also lead to the situation of the double-edged sword: recognition done by sufficiently credible nodes is credible, but not credible otherwise.

The repair by copying in information systems is also the double-edged sword and it is an engineering concern to identify when such repair can really eradicate abnormal elements from the system. This chapter considers cellular automata (CA), probabilistic CA specifically, to model the situation where computers in a network mutually repair by copying their content. We will focus on the simplicity rather than the reality of the system.

Studies on the reliability of information systems by a probabilistic CA are not new. For example, the results of Gacs (2001) for his self-simulating CA provided a counterexample to the one-dimensional CA version of the positive error rate conjecture.

11.3.2   Models of Probabilistic Cellular Automata

In our model, the system consists of nodes capable of repairing other connected nodes. We call the connected nodes "neighbor nodes" based on the terminology of CA. The repairing may be done by copying content to the other nodes, since we are considering application to networked computers.

Although mutual repairing and testing may be done in an asynchronous manner, our model considers synchronous interactions for simplicity and for comparison with existing probabilistic CA models (Domany and Kinzel 1984; Bagnoli et al. 1997; Vichniac et al. 1986).

The structure of the array is a ring with node 1 adjacent to node N (Fig. 11.2). Also, we restrict the case to each node having a binary state: normal (0) or abnormal (1).

The specific feature of the self-repair network is that repairing by an abnormal node has an adverse impact. This adverse impact is implemented as a higher probability of contamination than that of successful cleaning in the model by probabilistic cellular automata.

Each node tries to repair the nodes in the neighborhood in a synchronous fashion with a probability P_r. As shown in Fig. 10.1, the repairing will be successful with the probability P_{rm} when it is done by a normal node, but with the probability P_{ra} when it is done by an abnormal node ($P_{ra} \leqq P_{rm}$). In this chapter again, we assume $P_{rm} = 1$. The repaired nodes will be normal when all the repairing is successful. Thus, when repairing is done by two adjacent nodes, both of these repairs must be successful in order for the repaired node to be normal.

In such models, it is of interest to determine how the repairing probability P_r should be set when the success probability by abnormal node P_{ra} is given. Also, when P_r is fixed to some value and P_{ra} moves continuously to a large value, does the number of abnormal nodes change abruptly at some critical point or does it just gradually increase? Further, when given some value of P_{ra}, P_r should always be larger, which requires more cost.

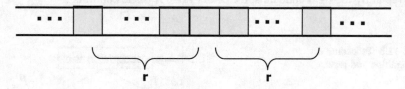

Fig. 11.2 One-dimensional array with the neighborhood radius r. The next state of the node (cell) will be determined by $2r + 1$ nodes in the neighborhood

The Domany-Kinzel model (Domany and Kinzel 1984) is a one-dimensional two-state and totalistic probabilistic cellular automaton (PCA). The interaction rule is as follows:

$$(0 * 0) \to 0 : 1, \ (0 * 1) \to 1 : p1, \ (1 * 1) \to 1 : p2.$$

Our PCA model can be equated with the DK model when $P_r = 1$ (i.e. nodes always repair).

11.3.3   Recognizing Abnormal Nodes Before Repair

Self-repair networks do not involve the recognition of abnormal nodes. Rather, the model is intended to show when and how the cleaning of the network can be done without recognition. The model in this chapter, however, focuses on the impact of recognition for diminishing useless repair or repair with adverse effect. Thus, the new model involves recognition of the states (normal or abnormal) of a target node before trying to repair the node. For simplicity, the frequency of recognition is controlled by a recognition rate P_{rec}. When recognition is done (with a probability P_{rec}), successful recognition occurs with a recognition success rate P_{recn} when done by normal nodes, and P_{reca} by abnormal nodes. If the target node is recognized as abnormal, repair action takes place. When recognition does not occur (with a probability $1 - P_{rec}$), the repair action takes place with the probability P_r. Thus, if the recognition is completely suppressed ($P_{rec} = 0$), this new model reduces back to the original model. Figure 11.3 shows the procedure of the recognition and repair.

Figure 11.4 depicts a scenario of the state change by recognition and repair. Considering the center node for example, recognition is done by the left node and the state of the center node is successfully identified. Since the center node is normal, no repair action takes place from the left node. On the other hand, recognition does not take place from the right node, and so repair action takes place in this scenario. Since the repair by the right node fails, the center node is made abnormal.

Fig. 11.3 Procedure for recognition and repair

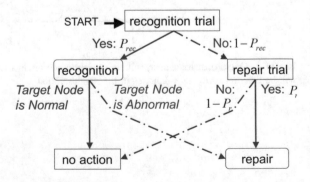

Fig. 11.4 An example of recognition and repair in the neighborhood

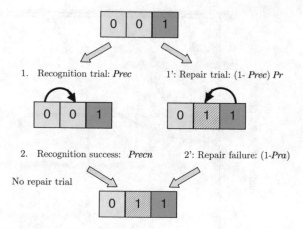

11.4   Simulation Results

11.4.1   Impact of Recognition

Computer simulations are conducted in a one-dimensional array with a ring structure (periodic boundary condition).

The parameters listed in Table 11.1 are fixed throughout the simulations. The other parameters P_{rec}, P_{reca}, P_r, and P_{ra} are varied to observe the impact of recognition.

We are concerned with the problem: Is recognition really necessary? If yes, when and how should the recognition be incorporated?

In the following simulations, we pursue the problem of identifying an appropriate level of recognition (i.e. P_{rec}) when the adverse effect of abnormal nodes (i.e. P_{reca} and P_{ra}) is given.

11.4.2   The Worst Case

Let us first consider the worst case when there will be no successful repair by abnormal nodes (i.e. $P_{ra} = 0$ is fixed).

Table 11.1 List of parameters for simulations

Symbol	Parameter	Value
N	Number of nodes	500
$N_{f(0)}$	Initial number of failure nodes	250
r	Radius of neighborhood	1
T	Time steps for each trial	5000
N_T	Number of trials averaged for one plot	10
P_{rn}	Repair success rate by normal nodes	10
P_{recn}	Recognition success rate by normal nodes	1

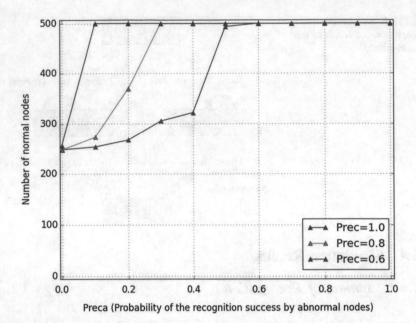

Fig. 11.5 Time evolution of the number of normal nodes when the recognition success rate by abnormal nodes P_{reca} is varied and $P_r = 0$, $P_{ra} = 0$ are fixed

Let us further divide the case into two extreme cases: there will be no repair action at all when recognition is not carried out (i.e. $P_r = 0$ is fixed), and there will always be repair when recognition is not carried out. Figure 11.5 plots the former case (*no repair*), and Fig. 11.6 the latter case (*always repair*). In the no-repair case, there is no blind repair and repair takes place only after recognition.

In the original model (Chap. 3 (Ishida 2005) and Chaps. 7–10) no recognition is involved, and so all the nodes will eventually become abnormal in the worst case. However, even the abnormal nodes can be eradicated in the model with recognition when P_{reca} exceeds 0.4. It is also observed that the level of recognition (P_{rec}) should be kept small to control the number of abnormal nodes when P_{reca} is less than 0.4.

In the case of *always repair*, both normal nodes and abnormal nodes can become extinct. There is a threshold value of P_{reca} between 0.9 and 0.8 above which abnormal nodes can be eradicated. Below the threshold, however, normal nodes will become extinct when P_{reca} is increased.

11.4.3   Level of Recognition Required for Eradication of Abnormal Nodes

When the rate of successful repair by abnormal nodes (i.e. P_{ra}) is given, what is the minimum level of recognition (i.e. P_{rec}) required for eradication of abnormal nodes?

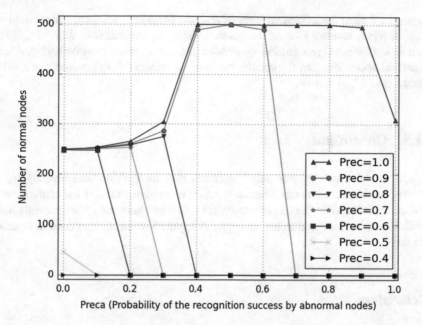

Fig. 11.6 Time evolution of the number of normal nodes when the recognition success rate by abnormal nodes P_{reca} is varied and $P_r = 1$, $P_{ra} = 0$ are fixed

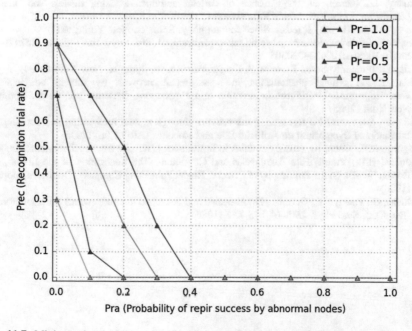

Fig. 11.7 Minimum level of P_{rec} to eradicate abnormal nodes when P_{ra} is given

Figure 11.7 plots the minimum level of P_{rec}. As observed, we do not care about the level of repair and the level of recognition when P_{ra} exceeds 0.4. When P_{ra} is less than 0.4, however, recognition is needed (P_{rec} becoming positive) to eradicate abnormal nodes. Further, the smaller the level of repair (P_r), the smaller the level of recognition (P_{rec}) can be.

11.5   Conclusion

A new model involving not only repair but also recognition was investigated to observe the impact of recognition. Only repair by copying suffices for eradication of abnormal nodes when the rate of successful repair by abnormal nodes exceeds some level. However, recognition before repair is required when the rate does not exceed the level.

References

Bagnoli, F., Boccara, N., Palmerini, P.: Phase transitions in a probabilistic cellular automaton with two absorbing states. http://www.arXiv.org.cond-mat/9705171 (1997)

Barabási, A.-L., Frangos, J.: Linked: The New Science of Networks Science of Networks. Basic Books, New York (2002)

Domany, E., Kinzel, W.: Equivalence of cellular automata to Ising models and directed percolation. Phys. Rev. Lett. **53**(4), 311–314 (1984)

Flower, D.D.R, Timmis, J. (eds.): Silico Immunology. Springer, New York (2007)

Gacs, P.: Reliable cellular automata with self-organization. J. Stat. Phys. **103**(1–2), 45–267 (2001). doi:10.1023/A:1004823720305

Ishida, Y.: Immunity-Based Systems: A Design Perspective. Springer, New York (2004)

Ishida, Y.: A critical phenomenon in a self-repair network by mutual copying. In: Knowledge-Based Intelligent Information and Engineering Systems, pp. 86-92. Springer, New York (2005)

Ishida, Y., Sugawara, Y.: Recovering the network through mutual recognition and copying. In: International Symposium on Artificial Life and Robotics (AROB'06) (2006)

Langman, R.E., Cohn, M. (eds.): Seminars in Immunology. http://www.idealibrary.com (2000)

Maturana, H.R., Varela, F.J.: Autopoiesis and Cognition—The Realization of the Living, ser. Boston Studies on the Philosophy of Science. Reidel Publishing Company, Dordrecht, Holland (1980)

Vichniac, G., Tamayo, P., Hartman, H.: Annealed and quenched inhomogeneous cellular automata (INCA). J. Stat. Phys. **45**(5–6), 875–883 (1986)

Chapter 12
Conclusion

We have examined the insightful question posed by von Neumann: Can system-level high reliability be attained with low-reliability components? We can answer affirmatively for a restricted version of the question by introducing autonomy as suggested by his insight on biological systems. That is, we need to introduce autonomy by allowing the components to act mutually: the self-action model. The self-action model creates a path from the self to itself, and emergent properties can be created as the feedback signal creates the possibility of stability (and instability as well).

The restricted version of the question using the autonomy exemplified by mutual repairing capability is a network cleaning problem: how can abnormal nodes be eradicated by mutual repair without node infection but with repair failure? Phase diagrams show that there is a region (the frozen region) where no abnormal node exists under the assumption that abnormal nodes are created only by repair failure. We also showed that even if it is impossible to eradicate all abnormal nodes, it is possible to contain them and prevent them from affecting the cluster of normal nodes by introducing strategic repairs in an evolutionary mechanism.

But this is not the end of the story, for we discussed the defense side only. Introducing autonomy also creates new dimensional threats. A key characteristic of the game theoretic approach is that both a player and an adversary are given the same options, and an endless battle begins like the battle (or boundary creation) between the self and nonself of the immune system.

We used a model to study the problem of cleaning a contaminated network by mutual copying, which turned out to be a generalization of the probabilistic cellular automata proposed by Domany and Kinzel. Cleaning by mutual copying is a double-edged sword, similar to the immune system, that could spread contamination rather than cleaning when not properly controlled.

When there is a threshold value in parameters, they should be carefully regulated. The framework for regulating mutual copying should ensure that nodes (corresponding to computers) do not become selfish by only being repaired, and also that nodes do not consume too many resources by frequent repairing. The framework involving a game theoretical approach of the spatial Prisoner's Dilemma turned out to be adaptive to the environment: strategies encouraging repairing

© Springer International Publishing Switzerland 2015
Y. Ishida, *Self-Repair Networks*, Intelligent Systems Reference Library 101,
DOI 10.1007/978-3-319-26447-9_12

emerge when the environment permits many abnormal nodes and the opposite strategies when the environment does not.

When the network is managed with autonomous distributed nodes using the game theoretic approach similar to the Prisoner's Dilemma, the adaptive character emerges. This adaptive nature is similar to the adaptive defense of the immune system that develops effective antibodies. Further, similarly to ecological systems, a symbiotic relation can be observed in the strategies that emerge. Similarities and differences from biological systems should be carefully investigated in a future work.

For complex and large-scale artificial systems such as the Internet, the systems tend to be out of control; central, planned control would be difficult to apply. For such systems, autonomous and distributed management is imperative and unavoidable rather than uniform control with a central authority. Autonomous and distributed management is favorable not only for control and management purposes but also for robustness against dynamic change, for such complex systems always undergo changes. When autonomous and distributed management is chosen as a framework, then we have to deal with selfish agents in exchange for leaving the control and management to each agent. The framework must guide the selfish actions of each agent toward the welfare of the entire system. Game theory has been studied and developed for such purpose, and has been applied to theoretical biology as well as economics.

The game theoretic approach to both microscopic and macroscopic models for the network cleaning problem in a self-repair network has been discussed. The game theoretic approach revealed conditions under which selfish agents can cooperate and form an organization of cooperative agents: their payoff must involve the survival of not only the agent itself but also its neighbors. By doing so, interacting agents can avoid not only being deadlocked waiting for neighbors' support (Nash equilibrium), but also being attracted to all the dead states (attractor).

Postscript: Challenges and Future

Grand Challenges

For computer systems including computer networks, separation of hardware design and software design was accomplished by surpassing the legacy of the *von Neumann architecture* with stored programs, but the separation is still in progress. For computer systems, *cloud computing* masks the users and even some programmers from hardware related affairs and troubles. For computer networks too, the *software defined network* (SDN) has attracted attention. As these trends continue, one important innovation for computer systems will be autonomy and self-management of computer systems. In other words, the high-performance information processing capability of computer systems should be turned on themselves as happens in most biological systems. This is the rationale for studying self-repair networks.

The three major names in this field, John von Neumann, Alan Turing and Norbert Wiener, laid the foundation of computer systems today, and a common inspiration and target of their work was biological systems.

Throughout this book we have often cited von Neumann's "Theory of self-reproducing automata" as well as "Probabilistic logics and the synthesis of reliable organisms from unreliable components." If he were to combine these theories with his other pioneering work on "Game theory and economic behavior," what kind of new science would emerge?

Meanwhile, Turing is famous not only for his Turing Machine, but also the Turing pattern and Turing model in morphogenesis: "The Chemical Basis of Morphogenesis" (Turing 1952).

Wiener is known for "Cybernetics: Control and communication in the animal and the machine" (Wiener 1948), and has had an influence on artificial systems

© Springer International Publishing Switzerland 2015   145
Y. Ishida, *Self-Repair Networks*, Intelligent Systems Reference Library 101,
DOI 10.1007/978-3-319-26447-9

including computer systems. The concept of homeostasis will become increasingly important for autonomous machines.

Regarding future challenges, if we look at more recent computer scientists and engineers, Jim Gray (Gray 2003) stated twelve Information-Technology Research Goals, among which the following three are consistent with self-repair networks:

9. Trouble-free system;
10. Secure system;
11. Always up.

Tony Hoare listed eight examples "not as recommendation but as examples that may be familiar from the past." Among the eight examples, "A mathematical model of the evolution of the web" (Hoare 2003) is related to self-repair networks. There is no doubt that self-repair networks as a model alone would not pass "the main tests for maturity and feasibility." But the topic may be worth pursuing in a more concrete form, for the web exhibits aspects of both a new artificial system and also a natural (collective) system devised by human beings.

Future Directions

After money was invented, the economic system of the free market developed its own dynamics and logics, and so cannot always be controlled externally from outside, leading to failures in monetary and financial systems. Similarly, after the Internet was invented, information systems have developed their own dynamics and logics. The Internet is evolving and expanding throughout human society. Although the Internet today may be out of control and difficult to redesign, we need to learn from its emergence as a large-scale information system to help us be more careful in designing and deploying information systems in the future, such as cloud computing systems, global sensor networks and the Internet of Things.

Over the last three decades, computers have become widespread through personalization and interconnection, and have had the following three main impacts:

• Global systemization,
• Local specialization,
• Glocal synchronization

For the autonomy of artificial systems, self-oriented computer systems should be addressed in both technology and theory. Regarding technology, self-controlled computation (focusing on the trade-off of power and accuracy, as well as resilience) has been studied as an extension from Green computing. Regarding the theory, studies on complex systems should focus on the autonomy of artificial systems. Autonomy comes with a huge cost and sophisticated mechanisms, possibly with several hierarchies. One challenge is how to decompose system-level autonomy to atomic-level autonomy or atomic agents. Also, as the mechanism design of game theory has struggled, it is necessary to design and manage an autonomous system composed of selfish agents to work in a stable fashion within the constraints of efficiency and fairness. *Strong AI* for managing the autonomous system needs to be revisited to address the fundamental *Frame Problem* in order for humans to be able to leave self-management tasks to machines that perform as expected, without having to worry that the machines may cause humans unexpected harm.

A Biological Design of Artificial Systems

For the design of information systems, we discussed a self-recognizing model for sensor-based systems and self-repair networks for actuator-based systems. Regarding the robustness and resilience of systems, one design principle is to have a uniform and yet developable omnipotent function unit (such as cells for biological systems). Even during the development phase, cells develop while referring to the environment even though the basic structure is framed by genetic information. The function of each component tends to *be degenerated* (not self-contained completely, requiring further information to be fully specified), and hence individuals reflect information about the environment.

For artificial information systems, it is still difficult to create such an omnipotent unit as cells. However, it is easier to rearrange the components of an information system than a mechanical system, for they are connected electronically rather than mechanically and so physical contact may not be necessary.

As a design paradigm for artificial systems, we have proposed a design that allows diversity in evolving in a dynamic environment learned from the immune system (Ishida 2004).

Simon emphasized the importance of hierarchical systems in design (Simon 1969). For the homeostasis of hierarchical systems, stabilization at each level of hierarchy may be required. For such degenerated and multi-layered stable system, we propose yet another model for the design: a matching automaton, which will rearrange and rematch existing components based on the affinity among them. This design can deal with hardware faults, for new types of threat are emerging such as power viruses and laser attacks that directly cause hardware faults rather than software faults that can be fixed by modifying the states.

Appendix

(Constants for the Steady-State of the Five Repair Types: Mutual AND, Mutual OR, Mixed AND, Mixed OR and Switching AND)

Tables A.1, A.2, A.3, A.4, A.5 and A.6.

Table A.1 Transition probability in each state transition (infection before repair)

State transition	Transition probability (mutual AND-repair)	Transition probability (mutual OR-repair)
(000) → 1	$P_r(1 - P_m)(2 - P_r + P_r P_m)$	$P_r(1 - P_m)(2 - P_r - P_r P_m)$
(001) → 1	$P_r^2(1 - P_m P_{ra}) + P_r(1 - P_r)((1 - P_m) + (1 - P_{ra})) +$ $P_i(1 - P_r)^2$	$-P_r^2(1 - P_m P_{ra}) + P_r((1 - P_m) + (1 - P_{ra})) +$ $P_i\{1 + P_r^2(1 - P_m P_{ra}) - P_r((1 - P_m) + (1 - P_{ra}))\}$
(101) → 1	$P_r(1 - P_{ra})(2 - P_r + P_r P_{ra}) +$ $P_i(2 - P_i)(1 - P_r)^2$	$P_r(1 - P_{ra})(2 - P_r - P_r P_{ra}) +$ $P_i(2 - P_i)\{1 - P_r(1 - P_{ra})(2 - P_r - P_r P_{ra})\}$
(010) → 1	$1 - P_r P_m(2(1 - P_r) + P_r P_m)$	$(1 - P_r)^2 + P_r(1 - P_m)(2 - P_r - P_r P_m)$
(011) → 1	$1 - P_r((P_m + P_{ra})(1 - P_r) + P_r P_{ra} P_m)$	$1 - P_r(P_m + P_{ra} - P_r P_{ra} P_m)$
(111) → 1	$1 - P_r P_{ra}(2(1 - P_r) + P_r P_{ra})$	$(1 - P_r)^2 + P_r(1 - P_{ra})(2 - P_r - P_r P_{ra})$

Table A.2 Coefficients of the equation expressed by parameters of the self-repair network (mutual repair)

Constant	Constants expressed by parameters (mutual AND-repair)	Constants expressed by parameters (mutual OR-repair)
A	$P_i^2(1 - P_r)^2$	$2P_i\{1 - P_r^2(P_r P_{ra} - 1) - P_r(2 - P_m - P_{ra})\} -$ $P_i(2 - P_i)\{1 - P_r(1 - P_{ra})(2 - P_r - P_r P_{ra})\}$
B	$-P_r^2(P_m - P_{ra})^2 - P_i(2 + P_i)(1 - P_r)^2$	$P_r^2(P_m - P_{ra})^2 + 4P_i\{1 - P_r^2(P_r P_{ra} - 1) - P_r(2 - P_m - P_{ra})\} +$ $P_i(2 - P_i)\{1 - P_r(1 - P_{ra})(2 - P_r - P_r P_{ra})\}$
C	$-2P_r(1 - P_m)(P_r(P_m - P_{ra}) + 1) +$ $P_r(P_r - 2P_{ra}) + 2P_i(1 - P_r)^2$	$P_r^2(1 + 2P_m P_{ra} - 2P_m^2) + 2P_r(-1 + P_m - P_{ra}) +$ $2P_i\{1 - P_r^2(P_r P_{ra} - 1) - P_r(2 - P_m - P_{ra})\}$
D	$P_r(1 - P_m)(2 - P_r + P_r P_m)$	$P_r(1 - P_m)(2 - P_r - P_r P_m)$

© Springer International Publishing Switzerland 2015
Y. Ishida, *Self-Repair Networks*, Intelligent Systems Reference Library 101,
DOI 10.1007/978-3-319-26447-9

Table A.3 Transition probability in each state transition where the following conventions are used to accentuate the AND-OR duality (Chap. 7)

	$Q_{rm} \equiv P_r P_m - P_r + 1$ $Q_{ra} \equiv P_r P_{ra} - P_r + 1$ $Q_r \equiv (1 - P_r)^3$	$R_{rm} \equiv 1 - P_r P_{rm}$ $R_{ra} \equiv 1 - P_r P_{ra}$ $R_r \equiv 1 - P_r$
State transition	Transition probability (mixed AND-repair)	Transition probability (mixed OR-repair)
(000) → 1	$1 - Q_{rm}^3$	$R_{rm}^3 - R_r^3$
(001)→1	$1 - Q_{rm}^2 Q_{ra} + P_i\{Q_{rm}Q_{ra}(Q_{rm} - Q_{ra}) + Q_r\}$	$R_{rm}^2 R_{ra} - R_r^3 + P_i\{R_r^3 - R_{rm}R_{ra}(R_{rm} - R_{ra})\}$
(101) → 1	$1 - Q_{rm}Q_{ra}^2 + P_i(2 - P_i)\{Q_{ra}^2(Q_{rm} - Q_{ra}) + Q_r\}$	$R_{rm}R_{ra}^2 - R_r^3 + P_i(2 - P_i)\{R_r^3 - R_{ra}^2(R_{rm} - R_{ra})\}$
(010)→1	$1 - Q_{rm}^2 Q_{ra} + Q_r$	$R_{rm}^2 R_{ra}$
(011) → 1	$1 - Q_{rm}Q_{ra}^2 + Q_r$	$R_{rm}R_{ra}^2$
(111) → 1	$1 - Q_{ra}^3 + Q_r$	R_{ra}^3

Table A.4 Coefficients of the equation expressed by parameters of the self-repair network (mixed repair) where the following conventions are used to accentuate the AND-OR duality (Chap. 7)

$Q_m \equiv P_r P_m - P_r + 1$
$Q_{ra} \equiv P_r P_{ra} - P_r + 1$
$Q_r \equiv (1 - P_r)^3$

$R_m \equiv 1 - P_r P_m$
$R_{ra} \equiv 1 - P_r P_{ra}$
$R_r \equiv 1 - P_r$

Constant	Constants expressed by parameters (mixed AND-repair)	Constants expressed by parameters (mixed OR-repair)
A	$(Q_m - Q_{ra})^3 + 2P_i Q_{ra}(Q_m - Q_{ra})^2 +$ $P_i^2\{Q_{ra}^2(Q_m - Q_{ra}) + Q_r\}$	$-(R_m - R_{ra})^3 - 2P_i R_{ra}(R_m - R_{ra})^2 +$ $P_i^2\{R_r^3 - R_{ra}^2(R_m - R_{ra})\}$
B	$-3Q_m(Q_m - Q_{ra})^2 + 2P_i\{Q_{ra}(Q_m - Q_{ra})(Q_{ra} - 2Q_m) - Q_r\} -$ $P_i^2\{Q_{ra}^2(Q_m - Q_{ra}) + Q_r\}$	$3R_m(R_m - R_{ra})^2 - 2P_i\{R_r^3 - R_{ra}(R_m - R_{ra})(2R_m - R_{ra})\} -$ $P_i^2\{R_r^3 - R_{ra}^2(R_m - R_{ra})\}$
C	$3Q_m{}^2(Q_m - Q_{ra}) + Q_r - 1$ $+ 2P_i\{Q_m Q_{ra}(Q_m - Q_{ra}) + Q_r\}$	$-3R_m^2(R_m - R_{ra}) + R_r^3 - 1+$ $2P_i\{R_r^3 - R_m R_{ra}(R_m - R_{ra})\}$
D	$1 - Q_m^3$	$R_m^3 - R_r^3$

Table A.5 Transition probability in each state transition

State transition	Transition probability (switching AND-repair)
$(000) \rightarrow 1$	$P_r(1-P_m)\{1-(1-P_{sr})P_r(1-P_m)\}\{P_{sr}+(1-P_{sr})(1-P_{sr}P_r+P_{sr}P_rP_m)\}+(1-P_{sr})P_r(1-P_m)$
$(001) \rightarrow 1$	$P_r\{1-(1-P_{sr})P_r(1-P_m)\}\{P_{sr}(1-P_m)+(1-P_{sr})(1-P_{ra})(1-P_{sr}P_r+P_{sr}P_rP_m)\}+(1-P_{sr})P_r(1-P_m)+$ $P_i\left[P_{sr}P_r(P_m-P_{ra})\{1-(1-P_{sr})P_r(1-P_m)\}\{1-(1-P_{sr})P_r(1-P_{ra})\}+(1-P_{sr}P_r)\{1-(1-P_{sr})P_r\}^2\right]$
$(101) \rightarrow 1$	$P_r\{1-(1-P_{sr})P_r(1-P_{ra})\}\{P_{sr}(1-P_m)+(1-P_{sr})(1-P_{ra})(1-P_{sr}P_r+P_{sr}P_rP_m)\}+(1-P_{sr})P_r(1-P_{ra})+$ $P_i(2-P_i)\left[P_{sr}P_r(P_m-P_{ra})\{1-(1-P_{sr})P_r(1-P_{ra})\}^2+(1-P_{sr}P_r)\{1-(1-P_{sr})P_r\}^2\right]$
$(010) \rightarrow 1$	$P_r\{1-(1-P_{sr})P_r(1-P_m)\}\{P_{sr}(1-P_m)+(1-P_{sr})(1-P_m)(1-P_{sr}P_r+P_{sr}P_rP_{ra})\}+$ $(1-P_{sr})P_r(1-P_m)+(1-P_{sr}P_r)\{1-(1-P_{sr})P_r\}^2$
$(011) \rightarrow 1$	$P_r(1-P_{ra})\{1-(1-P_{sr})P_r(1-P_m)\}\{P_{sr}+(1-P_{sr})(1-P_{sr}P_r+P_{sr}P_rP_{ra})\}+$ $(1-P_{sr})P_r(1-P_m)+(1-P_{sr}P_r)\{1-(1-P_{sr})P_r\}^2$
$(111) \rightarrow 1$	$P_r(1-P_{ra})\{1-(1-P_{sr})P_r(1-P_{ra})\}\{P_{sr}+(1-P_{sr})(1-P_{sr}P_r+P_{sr}P_rP_{ra})\}+$ $(1-P_{sr})P_r(1-P_{ra})+(1-P_{sr}P_r)\{1-(1-P_{sr})P_r\}^2$

Table A.6 Coefficients of the equation expressed by parameters of the self-repair network (switching AND-repair)

Constant	Constants expressed by parameters (switching AND-repair)
A	$P_{sr}(1-P_{sr})^2 P_r^3(P_m-P_{ra})^3 + 2P_i P_{sr}(1-P_{sr})P_r^2(P_m-P_{ra})^2\{1-((1-P_{sr})P_r(1-P_{ra})\}+$ $P_i^2 P_{sr}P_r(P_m-P_{ra})\{1-(1-P_{sr})P_r(1-P_{ra})\}^2 + P_i^2(1-P_{sr}P_r)\{1-(1-P_{sr})P_r\}^2$
B	$(1-P_{sr})P_r^2(P_m-P_{ra})^2\{3P_{sr}P_r(1-P_m)(1-P_{sr})-P_{sr}-1\}-$ $\quad 4P_i P_{sr}P_r(P_m-P_{ra})\{1-(1-P_{sr})P_r(1-P_m)\}\{1-(1-P_{sr})P_r(1-P_{ra})\}+$ $\quad\quad P_i(2-P_i)P_{sr}P_r(P_m-P_{ra})\{1-(1-P_{sr})P_r(1-P_{ra})\}^2 - P_i(2+P_i)(1-P_{sr}P_r)\{1-(1-P_{sr})P_r\}^2$
C	$P_{sr}^3 P_r^3\{3(1-P_m)^2(P_m-P_{ra})-1\} + P_{sr}^2 P_r^2[2\{(1-P_m)(P_m-P_{ra})(-3P_r(1-P_m)+1\}+2P_r-1]+$ $P_{sr}P_r[\{(P_m-P_{ra})^2(1-P_m)^2-1\}-P_r^2+1]+2P_r(P_m-P_{ra})(P_r P_m-P_r+1)+P_r(P_r-2)+$ $2P_i[P_{sr}P_r(P_m-P_{ra})\{1-(1-P_{sr})P_r(1-P_m)\}\{1-(1-P_{sr})P_r(1-P_{ra})\}+(1-P_{sr}P_r)\{1-(1-P_{sr})P_r\}^2]$
D	$P_r(1-P_m)\{1-(1-P_{sr})P_r(1-P_m)\}\{P_{sr}+(1-P_{sr})(1-P_{sr}P_r+P_{sr}P_r P_m)\}+$ $(1-P_{sr})P_r(1-P_m)$

Index

© Springer International Publishing Switzerland 2015
Y. Ishida, *Self-Repair Networks*, Intelligent Systems Reference Library 101,
DOI 10.1007/978-3-319-26447-9

Printed in the United States
By Bookmasters